TM 10-396

WAR DEPARTMENT TECHNICAL MANUAL

WAR DOGS

1943

WORLD WAR II

CIVILIAN REFERENCE EDITION

UNABRIDGED HISTORIC MILITARY HANDBOOK ON EMPLOYMENT AND TRAINING OF DOGS FOR WORK, WORKING BREEDS, ANIMAL CARE, AND TRAINING PROCEDURES

U.S. WAR DEPARTMENT, QUARTERMASTER CORPS

Doublebit Press

www.doublebitpress.com | Cherry, IL, USA

Original content under the public domain; unrestricted for civilian distribution.
Originally published in 1943 by the U.S. War Department, Quartermaster Corps.

A Part of the Military Outdoors Skills Series

Doublebit Press Civilian Reference Edition ISBNs
Hardcover: 978-164389-206-1
Paperback: 978-164389-207-8

TECHNICAL MANUAL No. 10-396 } WAR DEPARTMENT, WASHINGTON 25, D. C., 1 July 1943.

WAR DOGS

535335°—43——1

CHAPTER 1

GENERAL

SECTION I

MISSION, ORGANIZATION, AND ADMINISTRATION

1. Mission.—The Quartermaster General is charged with the responsibility of receiving, conditioning, training, and issuance of dogs for all war purposes, as well as of military personnel assigned to such activities. In executing this responsibility, The Quartermaster General is also authorized to develop doctrine concerning all phases of this work. As part of this mission, expert consultant service is provided to using agencies.

2. Organization.—*a.* Authority for all such functions is vested in the remount branch of the Office of The Quartermaster General. Responsible to the chief of remount are the reception and training centers established for training dogs and military personnel concerned. These centers are of two kinds:

(1) Existing remount installations.

(2) Independently established centers.

b. Direct responsibility for the proper handling, training, and care of war dogs at each unit to which these animals are assigned rests with the commanding officer of that unit. By continual inspection, supervision, and instruction, the commanding officer must insure that the provisions of the doctrine set forth in this manual are being carefully followed.

3. Operation of war dog reception and training centers.—*a. K-9 (Canine) section.*—The K-9 section at each center is administered through the administrative branch. Permanently assigned kennel men feed, guard, and care for all dogs as a regular routine.

Trainers may be required to assist in or even perform all duties in connection with kennel routine, as a matter of training. The administrative branch is also charged with the preparation and proper disposition of all necessary records and reports. These records and reports are to indicate any change in the status of war dogs, all necessary identification, and any remarks pertinent to the service of each dog. Reports are to be made whenever a dog is lost or stolen, whenever a dog dies, or whenever a dog appears unsuited to perform assigned duties.

b. Training branches of the K-9 section.—These branches are responsible for initial processing, classification, and training of both dogs and military personnel assigned to the center (fig. 1). The training includes—

(1) Basic training, which is fundamental to training for specialized functions work.

(2) Specialized training, which is designed to train dogs for one or another of the following functions:

(*a*) Interior guard duty—sentry.

(*b*) Interior guard duty—attack.

(*c*) Tactical service—messenger.

(*d*) Tactical service—scout.

(*e*) Tactical service—casualty.

c. Sledge and pack dogs fall in a separate category. (See FM 25-6.)

4. Acquisition and assignment of dogs.—For the present, dogs are procured through a patriotic organization, "Dogs for Defense, Inc." This organization obtains direct donations of dogs for war

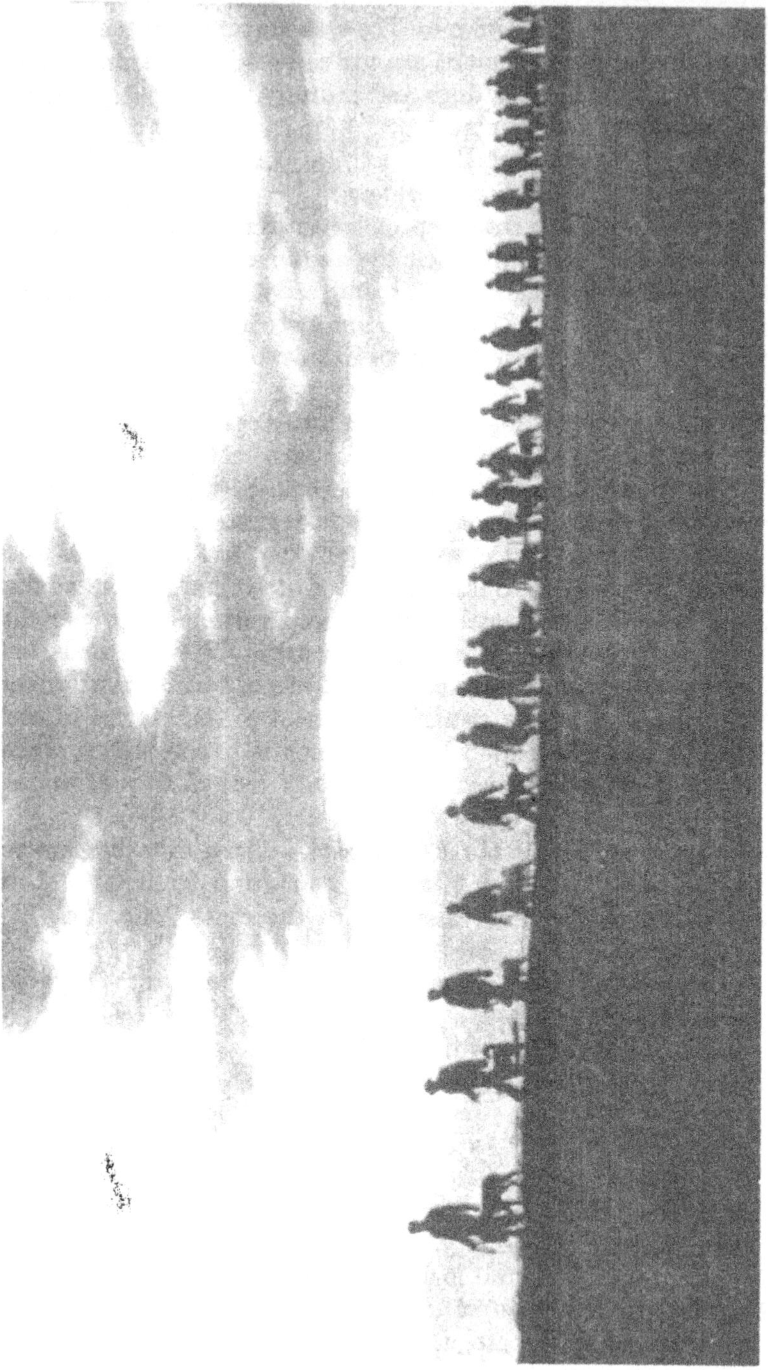

FIGURE 1.—Training war dogs.

purposes, and catalogs and initially screens suitable dogs for shipment to various war dog reception and training centers. Dogs trained for all purposes are issued to Army units and stations on approval requisitions that are cleared through the Office of the Chief of Remount.

5. Acquisition and assignment of personnel.—The station complement consists largely of quartermaster personnel concerned with administrative, training, and housekeeping functions. Units of the various arms and services seeking trained war dogs send men to a reception and training center where they receive necessary training. These men then return to their proper station with the war dogs they have trained.

SECTION II

HISTORY OF MILITARY USE OF DOGS

6. General.—The extraordinary characteristics of the dog—acuteness of his senses, his docility, his affection for man, his watchfulness, and his speed—enable him to be of great value for military purposes. This fact was recognized centuries ago. As methods of warfare changed through the ages, so did the military use of dogs change.

7. Prior to modern times.—*a.* The ancient Greek warriors made use of large dogs, thought by some to be the prototype of the modern mastiff, equipping them with heavy spiked collars similar to those sometimes used today; the Romans "drafted" the same species for attack work, recognizing them as a definite Army unit.

b. Later on, during the Middle Ages, war dogs often received the same complement of armor as did heavyweight chargers. They frequently were used to defend convoys.

c. During the Seven Years War, dogs were used as messengers by the army of Frederick the Great. Napoleon himself urged one of his generals to employ them as outposts in the Egyptian campaign at the end of the eighteenth century. Two centuries earlier, on this side of the Atlantic, they helped the Spaniards conquer Mexico and Peru.

d. It was the North American Indian who developed the use of dogs for pack and draft work as well as for sentry duty. Dogs were being used for these purposes long before the Spaniards introduced horses on the North and South American continents, which they settled in the sixteenth and seventeenth centuries.

8. From Russo-Japanese War to First World War.—*a.* Until the beginning of the present century most of the European powers had come to realize that dogs could be of value to their Armies. Ambulance dogs proved a success when tested out by the Russians during the Russo-Japanese War. The French were quick to utilize dogs as ammunition carriers when they found that the strong Pyrenean type could carry as many as 500 cartridges in a single load.

b. In desert and mountain warfare, dogs have played an equally important role as sentries and in helping to locate hidden gun posts. The Bulgarians conscripted their sheep dogs during the Balkan upheaval in 1910, and the Italians found it worth their while in 1911 to ship some of their shaggy, white-coated Maremma sheep dogs from the country around Rome across to Tripoli, where they were picketed in sand dugouts a few hundred yards ahead of the sentries to warn of enemy activity. The British, too, employed dog sentries, which served the purpose of the modern sound detectors, on the Abor Expedition in the Himalayas.

c. One of the interesting features of the long drawn-out Spanish-Morocco War was the use made of dogs by the Riffs. The animals were camouflaged in draperies which, in the hazy desert visibility, looked like a natives' burnous and made them indistinguishable from their owners. Trained to run along in front of the lines, they served to attract the Spanish fire and thereby reveal gun positions.

d. In Russia, the Imperial Army gave prominence to its dogs during a review held in May 1914, during which they were shown working under fire, carrying ammunition bags of 100 or more cartridges, and drawing miniature Maxims. They were also trained as couriers.

9. World War I.—*a.* It was during the First World War that dogs really made their mark. The Germans employed them in the greatest numbers, up to 30,000 messenger and ambulance dogs having served with their forces. The French used them for more varied work, and to what extent is shown by the fact that when the time came for demobilization in 1919 they had to dispose of more than 15,000. During the years 1914 to 1918 French military canine casualties exceeded 3,500 killed and 1,500 missing.

b. In this same conflict both the French and Belgian armies used draught dogs, and during the severe winter of 1915 some 400 sledge dogs, mostly Huskies, were brought from Canada. These operated in the deep snow that nearly brought to a halt all activities in the Vosges Mountains. It is recorded that one section of these dogs took but 4 days to haul 90 tons of ammunition up to a battery which men, horses, and mules had vainly struggled for a full fortnight to reach. When the snow melted the dogs were harnessed to a small-gage light railway and proved more useful than horses in that precipitous country. The cost of their "keep" was small, for there was an abundance of horse flesh, and a couple of teams of seven dogs each were able to do the work of five horses.

c. A major obstacle to movement during World War I was the heavy, sticky seas of mud. It was in traversing such terrain that the value of messenger dogs was demonstrated, for they could navigate it with comparative ease when men, up to their knees, were floundering and almost stalled.

d. Bloodhounds, too, were sometimes employed to prevent recently captured German prisoners from escaping back to their own lines, and the U. S. F. F. made continuous use of dogs as frontier guards after the revolutions.

10. World War II.—*a.* When the present war broke out the French army at once opened recruiting stations for army dogs.

b. For several years before the War, Germany had conducted war dog trials under a set of nationally uniform rules. Thousands of dogs were trained to serve as messenger and sentry dogs and in other capacities. During the fighting in France, the Germans were reported to have used many military dogs for patrol work.

c. For several years prior to the outbreak of war in the Pacific area, the Japanese had been transferring from German to Japanese registry, large numbers of dogs of types suitable for training, and there was evidence that these were trained for military service. The Japanese are known to have made considerable use of dogs in the Chinese theatre, having installed a training station at Nanking.

d. The British war office early in 1941 invited offers of dogs for military training. Some 7.000 dogs were brought forth and about one in three accepted for training. Success of the dogs in field trials resulted in an order that four dogs and two trainers be distributed per battalion throughout the British army. The tactical use of dogs by the British has been primarily along three lines:

(1) *Messenger dogs*, to carry messages between outposts and other missions in the rear or on the flank.

(2) *Patrol dogs*, to be led in advance of patrols, particularly at night, to pick up any humans to the front, and to indicate their approximate location by pointing.

(3) *Sentry dogs*, to be stationed at static positions such as forward machine-gun posts to indicate any advance against these positions.

e. The extensive program of the United States Army for the training of war dogs is described in section I.

CHAPTER 2

TRAITS AND CARE OF MILITARY DOGS

SECTION I

GENERAL

11. Scope.—*a.* In order to train war dogs effectively, it is necessary to know more than just principles of dog training. The effective trainer is one who—

(1) Has a sound understanding of the way in which a dog's mind works.

(2) Has a basic knowledge of breed traits.

(3) Is able through a knowledge of grooming, feeding, kenneling, and first aid to keep a dog in such physical condition as is necessary for performance with maximum efficiency.

b. This practical knowledge is essential to the student trainer of war dogs for the same reason that an enlisted man learning to operate a military motor vehicle is taught not only the techniques of operation but also nomenclature and function of vehicle parts and all important procedures of preventive maintenance. Consequently, information presented in chapter 2 must be made familiar to enlisted men who are being taught to train dogs and who later may serve as their masters in active service.

Section II

PSYCHOLOGY

12. General.—The dog's world differs from the human in some very specific ways. His world is predominately one of odors. His nose tells him countless things about the environment that entirely escape humans. He is more sensitive to sounds. His vision is considerably inferior to human vision, and for this reason he depends less upon it. He prefers to approach closely to objects that must be examined. However, his sensitivity to the movement of objects compares favorably with human sensitivity of this kind. To find a dog's ability or quality in a particular trait, one must seek it directly. To discover whether a dog is gun sure, one must test him with a gun. To discover whether he is intelligent and willing, one must train him. At present there is no reliable short cut.

13. Basic senses.—The senses of the dog with which the military trainer must concern himself chiefly are those of vision, hearing, smell, and touch.

a. Vision.—(1) *Structure and physiology of the eye.*—A most striking difference between the retina of the dog's eye and the human retina is that the former lacks a fovea. When a man focuses his eyes upon any object, the light reflected from that object is thrown upon the fovea of the retina. He can see many other objects besides that one, but they are seen indistinctly. This can be tested readily enough by focusing the eyes upon any word on this printed page and then trying to see how many other words can be read without moving the eyes. The words reflected upon the nonfoveal portions of the retina are blurred and poorly defined. Since the dog lacks a fovea, one may expect that even an object upon which he focuses is seen less clearly than it would be by humans. It seems certain that a dog can most conveniently and comfortably see objects which are at a distance of 20 feet or more.

(2) *Perception of movement.*—There is a type of visual stimulation to which dogs seem very sensitive. If any object is moved ever so

slightly, most dogs will detect and respond to the movement. This acuteness has been noticed in many psychological laboratories. Pavlov observed that the slight movement of an object in the vertical plane could be distinguished by dogs from movement in the horizontal plane, and also that discrimination between clockwise and counterclockwise motion was possible.

(3) *Color vision.*—The bulk of experimental evidence supports the opinion that to dogs the world looks like a black and white snapshot.

(4) *Value for training.*—Experimenters are agreed that dogs make scant use of their eyes in learning except in detection of motion.

b. Hearing.—(1) *Nature.*—Tests made in Russia and in Germany show that dogs hear sounds too faint to affect human ears. In one test, a German Shepherd at a distance of 24 meters responded to a sound which a man could not hear at a distance of more than 6 meters. Common observation supports these experimental findings. It seems apparent, too, that the dog hears sounds of higher pitch than affect human ears. The dog's ability to discriminate sounds of varying intensity is on a par with human ability. With respect to pitch discrimination, the case is not so clear.

(2) *Use in training.*—The exact elements of a command situation which are effective depend upon the nature of the dog's training. No doubt, inflection, the actual words and gesture all play a part. If it is intended to use a dog at night, or under any circumstances where the master cannot be seen, it is important that he be trained to respond properly without benefit of gesture. Most dogs can readily be instructed to respond to a number of oral commands. Some of them appear to understand most accurately the feeling of the master as it is conveyed by his voice. A word spoken in an encouraging tone will elate the dog. A cross word will depress it. Some dogs, however, cannot be reached effectively through the ear. These are generally not desirable for military training.

c. Smell.—(1) *Nature and physiology.*—Dogs so far surpass man in keenness of smell that it is difficult to imagine the nature of the sensations which they receive. Just as it is probably impossible for a dog to imagine what colors are, so it is impossible for the human to conceive of the vast range of odors and the delicate differences in chemical shadings to which dogs are so sensitive. The dog's nose is ideally adapted for the detection of minute amounts of odorous particles. Its snout is kept moist by a glandular secretion and is extremely sensitive to slight currents of air. Upon feeling such a current, the head is turned into the wind, the animal clears its nostrils and sniffs. A generous sample of the air passes into the nasal cavity and over the mucous membrane which is richly innervated with the finely subdivided endings of the olfactory nerves. This mucous membrane is supported on a complexly convoluted bony structure. Its structure is such as to present a maximum surface with a minimum obstruction to the circulation of the air. By comparison, the human olfactory apparatus is crude, yet even the human nose can detect chemicals borne by the air in such extreme dilution that they cannot be identified by the most sensitive chemical tests. In general, studies show that dogs can respond to odor traces of all known sorts and in dilutions far more extreme than can be detected by man. Furthermore, they can distinguish odors which seem identical to human beings, for example, natural and artificial musk.

(2) *Trailing.*—(*a*) Trailing capacity depends upon other traits besides olfactory acuity. Almost any Shepherd has sufficient acuity to become a good trailer, but only the exceptional dog possesses the ability to use his nose properly. More explicitly, it is only the exceptional dog which can be trained to follow slowly and carefully any trail his master sets him on, regardless of what other olfactory cues he may encounter. Any Shepherd can trail a rabbit, or a bitch in season. Only a few will, on command, follow the trail of a stranger leading through traffic, under snow, near a kennel full of dogs, or past a woodchuck hole.

(*b*) Tests indicate that the dog's success in trailing depends primarily upon his accurate discrimination of—

1. Earth odor from the compression and stronger vaporization of those spots stepped upon.

2. Plant odor from the destroyed vegetation.

3. Odor traces from shoes and shoe polish.

4. Odor traces from decaying animal or other organic matter.

5. Body odor specific to any particular person.

(*c*) Visual cues are of slight assistance to him.

(*d*) Many severe field trails supply good reason to believe that certain dogs can be trained to unravel a human trail despite numerous cross trails and other obstacles. Tests with chemical rather than natural trails lend further support to this conclusion.

(3) *Military significance.*—When working with dogs, a trainer can hardly avoid noticing that some are far more susceptible to odors than are others. For some forms of work, such as scouting, great acuity is essential. Many dogs with remarkable noses have been in the hands of efficient and observant masters who have employed them successfully and spectacularly in the apprehension of escaped prisoners under difficult circumstances. They have followed trails which were old, and crossed and recrossed by fresh ones; they have worked out trails covered by fresh snow. On the other hand, for some functions such as

casualty work, olfactory acuity is not essential to good work. In fact the dog's interest in odors may be brought to the attention of the trainer because it interferes with work. An animal may show more interest in the trails it crosses than in the work it is supposed to be doing. Except when especially required to do so by man, a dog does not frequently make use of his ability to follow a human trail. If he becomes separated from his master in unfamiliar surroundings he will trail him, but ordinarily, when using his nose for his own purposes, a dog pays attention chiefly to traces left by other dogs and by animals which are potential prey.

d. Touch.—There is a wide variation among dogs in the sense of touch. Certain dogs are very susceptible to manual caress or correction. Others appear to be relatively insensitive to it. These are generally not desirable for training.

14. Sensitivity.—In using the term "sensitivity" reference is made not so much to the stimulus threshold per se as to the threshold in terms of fear response. In other words, the oversensitive dog is startled by stimuli (sounds or touch) of lower insensity than is required to disturb the average dog, but his response is often one of flight and trembling. The normal dog responding to such stimuli might merely turn his head.

a. Relationship between body and ear sensitivity.—Records suggest that body and ear sensitivity vary quite independently. Of 123 dogs rated as undersensitive to touch, 51 were also undersensitive to sound and 67 were medium sensitive to sound. For 220 animals that were medium sensitive to touch, 140 were medium sensitive, 60 undersensitive and 20 oversensitive to sound. The indicated independence of the two forms of sensitivity suggests that shyness (oversensitivity) is not centrally determined, but is based upon receptor peculiarity. That is, it appears that shyness results from the extreme irritability of specific nerve endings; sureness, from the lack of it. A sound may actually "hurt" a gun-shy animal and yet a blow may not bother it.

b. Implications for military training.—Trainers should have no difficulty in rating dogs with respect to the efficacy of stimuli, and from a practical standpoint the classification is helpful. In training dogs, the voice and the hand are utilized almost exclusively in correction and reward. Thus, in the course of his regular work, a trainer cannot help but form a definite opinion with respect to the response of his pupil to auditory and tactual stimuli.

(1) *Oversensitive dogs.*—A dog that is oversensitive (shy) is so handicapped that it is not likely to demonstrate what intelligence it possesses in a form which the instructor can utilize. Dogs shy to either sound or touch are difficult to train and are unreliable. Certain

dogs showing only a mild degree of gun shyness can be accustomed to sound through repeated stimulation. Some dogs of this type, placed for several weeks in any Army fortification where there is constant target practice, can then be trained successfully for liaison work. But these near-shy dogs cannot be used where reliability in the face of noise is a life and death matter. In general, then, an oversensitive animal can be trained only with difficulty, if at all, and it cannot be trusted implicitly.

(2) *Undersensitive dogs.*—A dog that is undersensitive to both sound and touch is also very difficult to train. One cannot "reach" him readily to give either correction or caress. A dog sensitive to sound or touch, but not to both, can be instructed readily enough by a man who discovers the right approach. He must employ his voice in one case, his hands in the other.

(3) *Ideal dogs for training.*—The ideal dog (that is, ideal in the hands of a good trainer) is somewhat sensitive to both sound and touch. A mediocre teacher may spoil him. Such dogs tend to do very well or rather poorly, depending principally upon the wisdom with which they are handled.

(4) *General importance of trainer.*—It has been found that certain men, because of a lack of the proper range and timbre of voice, are unable to appeal to the dog successfully through his ear but, because of a certain finesse in muscular control, are excellent in handling the dog manually. The converse is also true. In general, each master succeeds best with a certain type of dog and each dog with a certain type of master. To secure maximum effectiveness, the qualifications of the man must be matched with the sensory peculiarities of the dog.

15. Energy.—This term refers to the degree of spontaneous activity of the dog; that is, to the speed and extent of his movements in general, not in response to any command. Dogs differ widely in degree of spontaneous activity exhibited, and the task of rating them in this respect is easier than that of rating for other functional traits. Above-average energy is not particularly necessary for military purposes.

16. Aggressiveness.—*a.* The extreme manifestation of aggressiveness is seen in attack. In general, a dog which is rated underaggressive cannot be taught to attack. The dog of average aggressiveness can be taught, though less readily, than an animal rated as overaggressive. The only difficulty in teaching the latter consists in securing prompt cessation of attack upon command.

b. The average and the underaggressive dogs rank about the same in energy, but the overaggressive group includes a high proportion of dogs of great energy. This is in line with the observed fact that a dog can be made mean and his aggressiveness increased by keeping him attached to a short chain for a couple of days. He cannot release the energy he generates. This method is employed in training for attack work those dogs which do not possess sufficient aggressiveness. In America, German Shepherds might not have earned a reputation for overaggressiveness had they not been kept so closely confined in small homes and in apartments.

17. Intelligence.—*a.* Intelligence generally is the trait most closely related to a dog's success in military training and service. In intelligence, the dog is far inferior to man but probably superior to any other animal below the primates. It is certain that neither in laboratory experiments nor in preparation for work have dogs had an opportunity fully to demonstrate their native intelligence. A dog can be taught to respond appropriately to an indefinitely large number of spoken words. No one knows the limit of his vocabulary. Under ordinary working conditions not more than a score of words are needed, but dogs have been known to master responses to well over 100 commands.

b. A dog's rating for intelligence is based upon the readiness with which he learns and the extent to which he retains and uses what he has learned. It should not be assumed that by intelligence is meant the ease with which an animal conforms to the demands of its trainer. Certain dogs learn very readily, but are instructed with great difficulty. They are able to learn how to avoid doing the work demanded by the trainer without, at the same time, incurring serious punishment. In a general way, then, a dog is rated high in intelligence if he is unusually capable of profiting by experience, regardless of whether this is to the full satisfaction of the trainer.

c. Some highly intelligent dogs are successful only when working with a man who "pleases" them. Under others they are unwilling and give the appearance of being stupid.

18. Willingness.—This term is an arbitrary one used to refer to the dog's reactions to man and *especially to his trainer or master.* The term applies not only to the nature of the dog's response in a command situation, particularly his response to an act which the animal has already learned, but also to learning of new duties. The dog may make the requisite response to the command or he may make some other response.

a. Willing dogs.—An animal is ranked high in willingness if he persistently responds to his master's requests with an effort to fulfill them, even though reward or correction is not immediate. Whether the animal possesses the requisite intelligence and physical strength, whether it succeeds or fails, is not considered.

b. Unwilling dogs.—If the promise of reward and the threat of punishment must constantly be before the dog in order that he work properly, he is considered an unwilling worker. There is a large group of dogs perfectly capable of executing the required movements but strongly inclined not to do so. Many such dogs appear to make a nice distinction between work and play. They will take great pleasure in retrieving, in searching for objects, in taking jumps. Such a dog, after training, will at times go to his master spontaneously and apparently suggest a romp which may include any of the acts mentioned. He will then respond, promptly and appropriately, to commands to retrieve, search, jump, trail, etc. If, however, the situation is reversed and the master initiates the activity, the dog may seem to have forgotten all he ever knew. The command FIND appears to be the same stimulus that had previously sent the dog eagerly seeking the hidden handkerchief, but now he responds with a cheerful but blank scrutiny of his master's face.

c. *Basic considerations.*—Certain considerations are basic to proper understanding of a dog's willingness or unwillingness.

(1) *Relation to trainer or master.*—A few dogs will work willingly for only certain people, often for only a single individual. Most dogs show unwillingness when commanded by a stranger. Certain animals have proved to be stubborn regardless of who trains and handles them. It is very evident that the best results are obtained when a dog and man are carefully matched. In any event, a dog's rating for willingness should be only in relation to his trainer or master. A dog should be penalized in his rating on this trait if he readily shifts his "willingness" to another man, a prisoner, for example.

(2) *Changes in willingness.*—Willingness can be enhanced or inhibited by the man who handles the dog. Improper handling may make a dog less willing at one time than at another. It is for this reason that some dogs may show less willingness under their trainers than under their masters during service. It is also for this reason that some dogs will work willingly during the first 5 minutes of a training period but, perhaps because of impatience of the trainer, unwillingly during the remainder of the period. Other dogs, however, just naturally seem to become unwilling after a few minutes. They should be rated relatively low in this trait.

(3) *Confusion with undersensitivity.*—Undersensitivity may be confused with stubbornness. Certainly a dog that is undersensitive to both sound and touch may appear to behave unwillingly, when as a matter of fact the commands and the motivation supplied by the instructor are less effective merely because of the limitations of the dog's sensory system.

(4) *Confusion with intelligence.*—When in doubt as to whether a dog's refusal to perform learned chores is due to stubbornness or forgetfulness, one can keep the animal shut in the kennels for a few days. The dog that is merely stubborn will then gladly run through his exercises for the privilege of being at large.

d. Military significance.—The trait of willingness is significant for several military functions.

(1) *Messenger and casualty work.*—In instruction for messenger and for casualty work the distinction between willing and unwilling dogs becomes especially clear. The dog usually works at a distance from, or out of sight of, his trainer. One type of dog will set off readily enough to establish a communication line or to seek a wounded man, but once well out of range of his trainer, he will conceal himself and remain carefully hidden, even though the trainer, seeking and shouting for him, passes within a few feet of his hiding place. Another dog will work diligently even though he has not received a command or encouragement for half an hour.

(2) *Guard work.*—Another form of work that clearly illustrates the effect upon the dog of remoteness of command or encouragement, is that of guarding a prisoner. The prisoner is informed that if he stands perfectly still the dog will not touch him but that if he takes a single step he will be attacked. There are records of dogs in actual service that have stood guard during the absence of their masters for 30 minutes and more. During these guard periods the master is entirely out of sight and is often at a great distance. The dog does not relax his vigil but is always ready to utter a warning growl and if necessary, to attack. He foils attempts of the prisoner to "inch" away to a distance from which he might safely run. He resists efforts of the prisoner to feed him or otherwise to win his friendship.

19. Motivation.—*a.* The trait that sets dogs apart from all other animals that have been studied experimentally by psychologists is their willingness to work for a reward of a most intangible nature—the approval of the experimenter. No rodents, cats, raccoons, monkeys, or apes have appeared anxious to please the scientists who strive so earnestly to study them. For these animals, the reward must be something of a practical nature, such as food, and the punishment, whether it is an electric shock or a slap, must be just as tangible. Canine subjects usually become attached to the experimenter who finds that a casual caress can be a remarkably effective reward, and a disappointed or disapproving word a potent punishment. Finally, even the anticipation of such disfavor clearly controls the dog's behavior.

b. Working dogs, even more than their laboratory brothers, provide opportunities to observe the efficacy of "intangible" reward and punishment. Once the master-dog relationship has been established, there is brought into play that motivation which finds its roots in the sentimental attachment of canine for man. Concrete punishment and reward are still used, and may be necessary on occasion, but to a large extent these may often be abandoned. It is more pleasant and more convenient to rely, so far as possible, upon the dog's eagerness to serve. Complete scientific explanation of this eagerness has not yet been attained. The important thing, however, is the actuality of this trait.

20. Sex differences.—The range of individual differences in each of the functional traits considered and in the quality of work produced is many times greater than any difference so far observed between averages for the two sexes.

Section III

BREEDS SUITABLE FOR MILITARY SERVICE

21. General.—Some breeds are generally more suitable for military service than others. However, regardless of breed, certain characteristics must be evident in all dogs selected for military service. These characteristics may be summarized under three headings.

a. Physical soundness.—Dogs for military use must have good bone, well-proportioned bodies, strong backs, deep chests, and strong muscular feet with hard, well-cushioned pads. Jaws must be strong and, except in the case of Boxers, teeth must be level. The nose should be large with well-opened nostrils. The forefeet should not turn markedly inward or outward and cowhocks should not be apparent.

b. Condition.—Dogs for military use should be easy keepers. They must never be gaunt or emaciated except under unavoidable field conditions and they must be muscular, never fat. Eyes should be clear and bright. Coats should show the bloom of health. The mucous membrane (eyes and gums) should be neither too dry not too wet, never pale. Breath should be sweet.

c. Behavior.—Dogs for military use should show a general alertness, vigor, and energy. They should be steady, not timid or excitable. They should show a willingness to be guided or taught, and evidence the ability to learn and to retain what they have learned. Crosses of the breeds described in the paragraphs below may be used for any military purpose for which they are qualified by virtue of their individual traits. (See chapter 4.)

22. Airedale Terrier.—*a. General appearance.*—The Airedale Terrier is the largest of the terriers; he is squarely built, standing about 23 inches high and weighing approximately 50 pounds. He has a hard, dense and wiry coat with typical black-and-tan markings.

b. Special traits.—Energy, endurance, ruggedness, speed, keenness, aggressiveness, and superior ability in water. The Airedale Terrier has been used extensively in foreign armies, especially the British.

23. Alaskan Malamute.—*a. General appearance.*—The Alaskan Malamute is a medium-sized dog with a strong, compact body, from 20 to 25 inches high, weighing between 50 to 85 pounds. He has a dense, thick, coarse coat, not too long, and usually colored wolfish gray or black and white. His face markings are a distinguishing feature, giving the appearance of a mask, which sets off the eyes. The eyes are almond-shaped and set obliquely, which gives them a wolflike appearance. The feet are especially distinctive; they are of the "snowshoe" type, large, long and flattish, with thick pads and abundant hair to cushion between the toes.

b. Special traits.—Strength, endurance, and tractability. His resistance to cold and his ability to move swiftly over snow and ice (because of the nature of his feet), together with his great strength and disposition to work, makes the Alaskan Malamute outstanding as a sledge dog.

24. Belgian Sheep Dog.

a. General appearance.—(1) There are two varieties of the Belgian Sheep Dog:

(*a*) The long-coated black.

(*b*) The short-coated or fawn-colored.

(2) The Belgian Sheep Dog is a strong alert dog of the Shepherd type, about 23 to 24 inches high, weighing from 50 to 55 pounds. The length of his body generally equals his height.

b. Special traits.—Strength, endurance, alertness, aggressiveness, and devotion. Belgian Sheep dogs were trained by the thousands during World War I to carry messages between outflung sectors. Many gave their lives.

25. Bouvier de Flandres.—*a. General appearance.*—The Bouvier is a dog of powerful build, rugged looking, from 23 to 28 inches high, weighing between 70 and 100 pounds. His coat is rough and harsh and very thick; so rough that it always appears to be badly groomed. There is a thick wooly undercoat for winter protection. His color may be any shade from fawn to black, including brindle and pepper-and-salt. A heavy, fairly long, rough beard and moustache are characteristics of this breed.

b. Special traits.—Strength, energy, endurance, courage, and tractability. The Bouvier has been used with great success in foreign armies, notably the Belgian Army in the last war.

26. Boxer.—*a. General appearance.*—The Boxer is a medium-sized, smooth-haired, sturdy dog of short, square figure and strong limbs, between 21 and 24 inches high, weighing between 65 and 80 pounds. His coat is short and shiny, lying close to the body. It is fawn or brindle in color, with a black mask. The head is the distinctive feature of the breed; it is strong and square with a distinct stop, somewhat suggestive of a Bulldog, but with a longer, stronger muzzle and narrower skull. The Boxer is the only breed suitable for Army use in which the undershot jaw is allowed.

b. Special traits.—Aggressiveness, strength, tractability, and natural distrust of strangers.

27. Briard.—*a. General appearance.*—The Briard is a strong and substantially built dog, lithe, muscular, and active. He is between 22 and 27 inches high, weighing between 65 and 80 pounds. His coat is long, slightly wavy, but stiff and strong, and of any color except white, being either solid, brindle, or pepper-and-salt.

b. Special traits.—Courage, strength, water-resisting coat, tractability, and acute hearing. In the last war Briards were used for messenger work and acted as sentries at advance posts. They accompanied patrols and carried food and supplies and even munitions to the front line, and so many of them died in the service that the breed was greatly reduced in numbers. Briards were also used in France for drawing carts.

28. Bull Mastiff.—*a. General appearance.*—The Bull Mastiff is a large dog of powerful symmetrical build. He is between 24 and 27 inches high and weighs between 100 and 115 pounds. His coat is short and dense, of any shade of fawn or brindle.

b. Special traits.—Strength, tractability, and keen nose.

29. Chesapeake Bay Dog.—*a. General appearance.*—The Chesapeake Bay Dog is a rugged, sturdy animal between 21 and 26 inches high, weighing 60 to 75 pounds. His coat is thick, short, harsh, and oily. He has a dense wooly undercoat. His color is any variation of brown, from dark brown to faded tan.

b. Special traits.—Strength, endurance, keen nose, and exceptional ability as a swimmer. The oil in the outer coat and the wooly under coat are of great value in preventing cold and water from reaching the dog's skin.

30. Collie.—*a. General appearance.*—There are two varieties; the rough Collie and the smooth Collie. Except for the difference in their coats they are very much alike; lithe, active dogs, between 22 and 25 inches high, weighing from 50 to 60 pounds. The rough Collie has a coat that is abundant except on the head and the legs. The outer layer is harsh, straight, and thick; the inner one soft and furry. The color can be any combination of black and tan, with white collar, or sable with white markings. The smooth Collie has a short coat.

b. Special traits.—Speed, alertness, endurance, and tractability.

31. Curly-coated Retriever.—*a. General appearance.*—The Curley-coated Retriever is a dog of medium size, from 19 to 22 inches high, weighing between 40 and 55 pounds. His coat is a mass of crisp curls, black or liver-colored.

b. Special traits.—Ability to resist and stand severe weather, and endurance.

32. Dalmatian.—*a. General appearance.*—The distinctive feature of the Dalmatian is his slick, short, white coat decorated with round, black spots or deep brown ones. He is muscular and active, between 20 and 23 inches high, weighing between 40 and 55 pounds. His legs and feet are of great importance since he is used for covering long distances behind horses or horse-drawn vehicles.

b. Special traits.—Endurance, liking for work, tractability, poise, and ability to cover long distances. Dalmatians must be carefully tested for hearing before being accepted for military use, as occasionally members of this breed are found to be deaf. He is best used in those situations where his flashy appearance does not constitute too great a hazard of detection. He can be colored to harmonize with his surroundings. The British usually paint him shades of brown.

33. Doberman-Pinscher.—*a. General appearance.*—The Doberman-Pinscher is a compact, muscular dog, from 24 to 28 inches high, weighing between 50 and 70 pounds. His coat is short, hard, thick, and close-lying, usually black with typical black-and-tan markings.

b. Special traits.—Nervous energy, speed, power, keen nose, tractability, and exceptional agility. The Doberman-Pinscher has been used as a police and war dog with great success. His short coat makes him best suited for temperate climates.

34. English Springer Spaniel.—*a. General appearance.*—The English Springer Spaniel is a medium-sized variety of spaniel, active and strong. He is from 19 to 22 inches high and weighs between 42 and 55 pounds. His coat is flat or wavy, of medium length, and dense enough to be waterproof. It is colored liver and white, black and white, tan, or roan.

b. Special traits.—Speed, keen nose, and willingness to work.

35. Eskimo.—*a. General appearance.*—The Eskimo is a compact, sturdy dog capable of heavy hauling and other hard work. He is from 20 to 25 inches high, weighing 50 to 85 pounds. His coat has a heavy covering of coarse hair with a dense woolly under coat. It is black, white, gray, tan, buff, or some combination of these colors. His "snowshoe" feet are a distinctive characteristic of the breed.

b. Special traits.—Strength, endurance, power, keen nose, memory for routes, and ability to work as a member of a team. Because of their "snowshoe" feet, these dogs can haul loads for miles over rough ice and crusted snow without getting footsore. An Eskimo team will draw from one and a half to double its body weight, and average from 20 to 30 miles daily on long trips.

36. Flat-coated Retriever.—*a. General appearance.*—The Flat-coated Retriever is an active dog of medium size, between 20 and 23 inches high, weighing from 60 to 70 pounds. He has a dense, flat coat, either black or liver-colored.

b. Special traits.—Superior ability in swimming and endurance.

37. German Shepherd Dog.—*a. General appearance.*—The German Shepherd is a working dog of wolflike appearance. He is of good middle size and muscular body. He has a double coat to protect him in all kinds of weather; a harsh, straight outer coat of medium length, and a dense, woolly undercoat. In color, he may run from black to white; the most characteristic color is wolf-gray.

b. Special traits.—Keen nose, endurance, reliability, speed, power, tractability, and courage. German Shepherds have been used successfully as police and military dogs.

38. German Short-haired Pointer.—*a. General appearance.*—The German Short-haired Pointer is a lithe and muscular dog from 22 to 25 inches high, weighing from 50 to 70 pounds. His coat is short, flat, and firm and solid liver, or liver and white, ticked or spotted. Liver color is an outstanding feature of this breed.

b. Special traits.—Speed, stamina, keen nose (especially at night), and alertness. He is usually a good swimmer.

39. Giant Schnauzer.—*a. General appearance.*—The Giant Schnauzer is a robust, sinewy, dog, built on terrior lines, rather heavily set and squarely built. His coat is close, strong, hard and wiry, colored black or black-and-tan, brindle or pepper-and-salt. He is between 22 and 26 inches high, weighing between 65 and 100 pounds. He has heavy, stubby whiskers.

b. Special traits.—Reliability, energy, power, tractability, keen nose, speed, liking for work, ruggedness, aggressiveness, and ability to stand extreme weather. The Giant Schnauzer has been used successfully for the past 30 years as a police and military dog in foreign countries.

40. Great Dane.—*a. General appearance.*—The Great Dane is the tallest dog suitable for military duties. He stands not less than 30 inches high, and weighs between 100 and 150 pounds. His coat is very short, thick, smooth, and glossy, colored black fawn, brindle, or sometimes white with black patches irregularly distributed over the entire body. This latter variation is known as harlequin.

b. Special traits.—Power, strength, tractability, and poise. The Great Dane does not usually exhibit nervousness even under conditions of great excitement.

41. Great Pyrenees.—*a. General appearance.*—The Great Pyrenees is a dog of great size and strength. He stands from 27 to 32 inches high and weighs between 100 and 125 pounds. His coat is of medium length, with a heavy, fine white under coat and long, flat, thick, outer coat of coarser hair, usually all white or white with grey or tan markings.

b. Special traits.—Great strength, and ability to withstand severe weather.

42. Irish Setter.—*a. General appearance.*—The Irish Setter is a spirited, medium-sized dog of the setter type, from 21 to 24 inches high, weighing between 50 and 60 pounds. His coat is of moderate length, flat, and colored a rich chestnut or mahogany brown.

b. Special traits.—Keen nose. speed, courage, and stamina.

43. Irish Water Spaniel.—*a. General appearance.*—The Irish Water Spaniel is a strongly built dog of rugged appearance. He is between 21 and 24 inches high, and weighs between 50 and 65 pounds. His coat, always of a solid liver color, is curled in tight, crisp ringlets. He has a so-called "rat tail."

b. Special traits.—Keen nose, water-resisting coat, and ability to swim well. He is fairly aggressive.

44. Labrador Retriever.—*a. General appearance.*—The Labrador Retriever is a strongly built, very active dog, from 21 to 25 inches high, weighing between 55 and 70 pounds. His coat is usually black, although golden or wheaten specimens occur. It is fairly short, very dense, straight, and quite hard. His tail, a distinctive feature of the breed, is very thick toward the base, gradually tapering toward the tip.

b. Special traits.—Keen nose, weather-resisting coat, superior ability in swimming, strength, and endurance; he is not particularly aggressive.

45. Newfoundland.—*a. General appearance.*—The Newfoundland is a massive, powerful dog, standing between 26 and 30 inches high, and weighing between 120 and 150 pounds. His coat is dense, of medium length, coarse and oily, and water-resistant. He is usually jet black, but white and black is a common variation.

b. Special traits.—Great strength, superior ability in swimming, and ability to endure cold and wet weather. The Newfoundland has earned a reputation as a rescuer of persons in danger of drowning.

46. Norwegian Elkhound.—*a. General appearance.*—The Norwegian Elkhound is a typical northern dog, of medium size, with a compact, proportionately short body, from 18 to 21 inches high, weighing between 50 and 55 pounds. His coat is thick and rich, but not bristling, and gray in color.

b. Special traits.—Speed, endurance, tractability, and aggressiveness.

47. Pointer.—*a. General appearance.*—The Pointer is a lithe and muscular dog, from 23 to 26 inches high, weighing from 50 to 70 pounds. His coat is short and close. His color is usually white with rich liver markings; lemon and white, orange and white, black and white, and sometimes even solid black, or other colorings.

b. Special traits.—Energy, willingness to work, and keen nose.

48. Rottweiler.—*a. General appearance.*—The Rottweiler is a strongly built, active dog between 22 and 27 inches high, weighing between 70 and 100 pounds. His coat is short, coarse, and flat, and colored black with typical black and tan markings.

b. Special traits.—Strength, courage, poise, dependability, and aggressiveness.

49. St. Bernard.—*a. General appearance.*—The St. Bernard is a massive, powerful, upstanding dog, strong and muscular, with a powerful, heavy head. He stands from 27 to 30 inches high, weighing from 125 to 150 pounds, with a dense coat of medium length, colored red or light brindle with white markings.

b. Special traits.—Great strength, good nose, and ability to endure cold and wet weather. The St. Bernard is famous as a "hospice" dog (or rescue dog), so called because he was used by the monks of the hospice of St. Bernard in the Swiss Alps. These monks trained their dogs to rescue travelers lost in the snow.

50. Samoyede.—*a. General appearance.*—The Samoyede is strong and active, with a heavy, weather-resisting coat. He is from 18 to 22 inches high, weighing 40 to 55 pounds. His coat consists of a thick, harsh, straight outer coat with a thick soft under coat lying close to the body; its color is some shade of white or cream. He has "snowshoe" feet.

b. Special traits.—Strength, ability to work in a team, endurance, ability to withstand cold weather, and feet suitable for travel over snow and ice.

51. Siberian Husky.—*a. General appearance.*—The Siberian Husky is an active dog, quick on his feet. He is from 20 to 24 inches high, and weighs from 40 to 60 pounds. He has a strong, moderately compact body; his coat is thick but soft, with a dense under coat. It is usually of medium length. All markings are found in this breed; the most usual colors are various shades of wolf-and-silver-gray, or black with white points. He has "snowshoe" feet of medium size, well furred between tough, thickly cushioned pads.

b. Special traits.—Speed, endurance, ability to withstand cold, ability to work in a team, and tractability.

52. Standard Poodle.—*a. General appearance.*—The Standard Poodle as a military dog presents a far different appearance from the traditionally landscaped poodle seen at dog shows. He is clipped all over for Army work, and his coat allowed to grow out to a length of 1 or 2 inches, either all over, or with the face and feet clipped bare. Thus cut down, the poodle looks like a medium-sized retriever. He stands from 20 to 25 inches high, weighing from 50 to 75 pounds. His coat is tightly curled, very dense, of any solid color. He is a sturdy, squarely built dog, active and poised.

b. Special traits.—Unusual ability to learn rapidly, good retention, patience, agility, versatility, courage, keen nose and hearing. The poodle's one drawback is his rapidly growing coat, which is never shed

FIGURE 2.—Reception and training center.

and must constantly be cut down to prevent its becoming matted or knotted, or entangled with foreign matter. Some specimens are too small or too light to be serviceable.

53. Wire-haired Pointing Griffon.—*a. General appearance.*—The Wire-haired Pointing Griffon is a rugged, hardy dog of the pointer type, between 20 and 23 inches tall, weighing from 50 to 60 pounds. He is fairly short-backed, rather low on his legs, and strongly limbed. His coat, a grizzled gray, chestnut splashed, is harsh like the bristles of a wild boar.

b. Special traits.—Keen nose, stamina, vigor, and speed.

Section IV

SHIPMENT AND RECEPTION OF NEW DOGS

54. Manner of shipment.—*a.* All prospective war dogs are shipped by express in individual shipping crates (fig. 3) provided by

FIGURE 3.—Shipping crate.

the Government. These crates are marked with shipping tags which carry the following information: the name, the D. F. D. (Dogs for Defense) number assigned him, and the breed of the dog.

b. Most shipments which comprise more than 10 dogs are accompanied by an enlisted man or noncommissioned officer who has received the dogs at their original shipping point.

FIGURE 4.—Receiving a dog at a war dog reception and training center.

55. Reception procedure.—When a dog is taken from his shipping crate (fig. 4) he is immediately equipped with a leather collar carrying a metal tag bearing the D. F. D. number assigned to this dog. After being exercised on leash and permitted to evacuate, the dog is put in quarantine in an individual compartment of the multiple-unit reception kennel under the supervision of the veterinarian. There he is provided with food and water and left alone for a period long enough to enable him to recover from his journey and become accustomed to his surroundings.

56. Recording procedures.—On the day following the dog's arrival, data are obtained for the Record Card (W. D., Q. M. C. Form No. 120) (fig. 5). Included are the dog's weight, his height from shoulder to the ground, a general description of color, breed, sex, and official number.

57. Tattooing procedure.—*a.* After all necessary information has been obtained for the record, the dog is taken to the veterinarian for tattooing of the dog's number on the left ear, on the flank, or on the belly. This is required for purposes of identification.

FIGURE 5.—Examining a new arrival.

b. Using the Preston system, it is possible to tattoo 4,000 animals with each letter assigned, in the following manner: if the letter "A" is to be used, the first animal tattooed will receive the tattoo "A000", the second "A001", the third "A002" and so on up to "A999", which makes the first thousand. The second thousand animals will be tattooed "0A00", "0A01", "0A02" and so on up to "9A99." The third thousand will be tattooed "00A0", "00A1", "00A2" and so on up to "99A9". The fourth thousand will be tattooed "000A", "001A", "002A", and so on up to "999A."

SECTION V

GROOMING AND CARE

58. General.—With dogs as with horses, a daily grooming is necessary the year round. If a dog is groomed properly every day, it will seldom be necessary to bathe him. A good brushing will keep a dog clean and maintain his coat and skin in healthy condition. The dog must be kept free from fleas, lice, and ticks. Grooming includes inspection and care of ears, nails, eyes, nose, teeth, and anal gland. A dog's appearance denotes his state of health as well as the care he receives, and reflects directly on his handler.

59. Coat.—Daily brushing is important (fig. 6). A dog's coat is a special development of the skin where each hair grows from a separate hair follicle. It was developed as a protection from rain, excessive heat, or cold. Some breeds have a double coat; the deeper layer, or under coat, composed of soft, wooly hair and the outer layer, or outer coat, composed of more or less coarse, stiff hair, somewhat oily and water-resistant. A good brisk rub-down with the fingertips loosens up the dead skin. A thorough but gentle brushing, following the rub-down, not only keeps the dog's coat clean and free from foreign matter but also polishes and burnishes the coat and imparts a healthy glowing appearance, or bloom. The coat should be brushed first against and then with the grain. In winter, when the dog's undercoat is of great importance, combing should be limited, as excessive combing will tear out the warm under coat, thus leaving the dog more exposed to the weather.

60. Nails.—*The daily grooming will include inspection of the dog's nails* (fig. 7). Long nails often break or grow into the pads of the feet, rendering a dog unfit for service. Particular attention should be given to the first digits on the front and hind feet, which compare

FIGURE 6.—The daily brushing.

FIGURE 7.—Care of the nails.

to our thumbs and large toes. In dogs, these digits do not come in contact with the ground and therefore do not have a wearing surface. Some dogs have an additional toe on the inside of the foot adjacent to the thumbs or big toes. This digit as well as the first digit on the back feet are known as dew claws. They serve no useful purpose to the dog. They may be removed surgically by the *veterinary officer.*

61. Eyes.—*Eyes must be given attention.* The eyes are bathed daily with a warm solution of boric acid. Should the dog evidence a mucous discharge from the eyes, the dog should be examined by the veterinarian, as eyes are referred to as the mirror of the body and their appearance will often indicate the onset of infections.

62. Ears.—*a. Grooming will always include daily cleaning of the ear flaps and ear canals.* The ear flap is brushed and all matted dirt and food removed. The flap is then examined for wounds, thickening of the margins, and other abnormal conditions.

b. The ear flaps were designed by nature as a protective cover for the ear against the entrance of foreign matter. However, dogs with long overhanging ear flaps, such as poodles, retrievers, and spaniels, accumulate a moist, thick brown wax in the ear canal, caused by a lack of sufficient air to dry the ear canal. The ear should be examined daily and all visible wax removed. This can be done with a piece of cotton, dipped in a boric acid solution or a solution of equal parts of hydrogen peroxide and water and applied to the visible parts of the ear canal in the inside flap of the ear. The operation should be performed in a gentle manner and the swab must not be forced into the ear canal. As many swabs should be used as are necessary to thoroughly clean the ears.

c. When the wax has been removed, the ear canal should be examined for excessive hair growth. In some breeds of dogs, notably poodles, this condition is very common and, unless the hair is removed at frequent intervals, it will become matted and cause the dog considerable discomfort.

d. Dogs evidencing symptoms of ear trouble will constantly shake their head, twitch their ears, and scratch their ears with their hind feet. When these symptoms persist, the dogs should be presented to the veterinarian for examination and treatment.

63. Nose.—The nose must be inspected for cuts, scratches, or mucous discharge. Cuts and scratches should be kept clean. In the case of a persistent watery or thick discharge the veterinarian should be consulted. An excessively dry or moist nose is sometimes a symptom of ill health.

64. Teeth.—Teeth must be inspected. They require cleaning when there is an accumulation of tartar. Sometimes dogs have abnormal or diseased teeth that must be removed.

65. Anal glands.—In the care of dogs, it should be kept in mind that the anal glands, which are small glands situated on either side of the anus or rectum, often become infected or impacted, causing severe pain or annoyance to the animal. These glands should be expressed by placing a large piece of cotton over the anus and pressing firmly with the thumb and fingers on both sides of the rectum, expelling the impacted matter into the cotton. Infected glands require treatment by the veterinarian.

66. Skin parasites.—*a.* Fleas, lice, and ticks are common insect pests. Fleas not only cause the dog great annoyance, but at times are the cause of eczema. They are also the intermediate host of the dog tapeworm. The eggs of the flea do not remain attached to the coat of the dog, but fall into the bedding or cracks in the floor where they hatch and undergo part of their life cycle. Consequently, the elimination of fleas from the animal will do little to correct this condition unless the bed and kennel are also disinfected.

b. Dog lice are small sucking and biting insects which attach themselves to the dog's body. The eggs of the lice, unlike those of the flea, are attached to the hair and may not be affected by the agents used to kill the adult lice. To eliminate lice, strict sanitation, careful grooming, and the repeated application of parasiticides are necessary. The bedding should be burned and kennels thoroughly disinfected. Ticks should be removed carefully by means of tweezers, and should not come in contact with the hands, as some ticks are carriers of diseases transmissible to man. A small amount of ether on cotton placed over the tick prior to removal will cause it to remove its head from the animal's skin. This facilitates removal of the entire tick. Daily grooming will help keep the dogs free from skin parasites, which may be eliminated by dipping, spraying, or the use of insect powder as prescribed by the veterinarians.

67. Bathing.—A normal healthy dog should be bathed as infrequently as possible.

a. The skin of the dog is peculiar. It is quite rich in grease glands and deficient in sweat glands. This oil keeps the skin soft and prevents its becoming dry and cracking. It also protects the coat and makes the outer coat water-resistant.

b. When a dog is bathed frequently, the natural oil is removed from the skin and the skin and coat become unnaturally dry, resulting in a poor coat and minute cracks in the skin. These cracks cause an irritation which makes the dog scratch and bite himself, thus preparing the way for eczema or other infectious skin ailments.

c. When a bath is necessary, the dog's eyes and ears must be protected from soap and water before he is put into the tub. This is accomplished by squeezing a small quantity of yellow oxide of mercury or a drop of castor oil into each eye. Cotton must be put into each ear. The water used for bathing a dog should be warm, about 105° Fahrenheit, never hot or cold. Any good soap which does not contain too much free alkali may be used, provided care is taken to rinse out all traces of soap before the dog is removed from the bath. Soap, when left in the coat, becomes sticky and collects dirt. It also

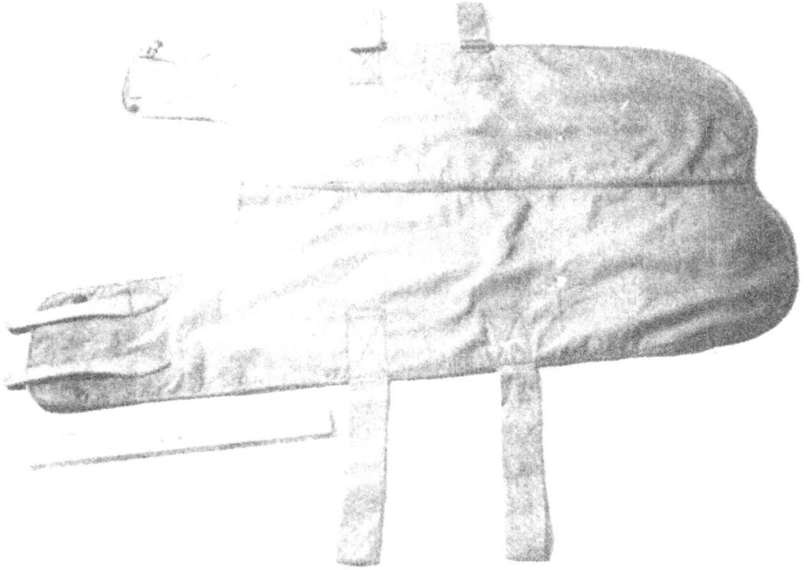

FIGURE 8.—Waterproof blanket.

causes an irritation of the skin. The dog should be dried with towels as thoroughly as possible, and then massaged by hand and brushed. Finally, as an aid to drying, he may be encouraged to run outdoors for a while, preferably if the sun is shining, never when the weather is very cold.

68. Waterproof blankets.—These are provided for use in wet weather (figs. 8 and 9).

69. Trimming and clipping.—Certain breeds, such as Airedales, Giant Schnauzers, and other terrierlike wire-coated dogs, must be trimmed to keep them looking presentable. Poodles must frequently be clipped or cut down because their coats grow very rapidly.

70. Evacuation of intestines and bladder.—Dogs must be allowed to empty both bladder and intestines at regular intervals.

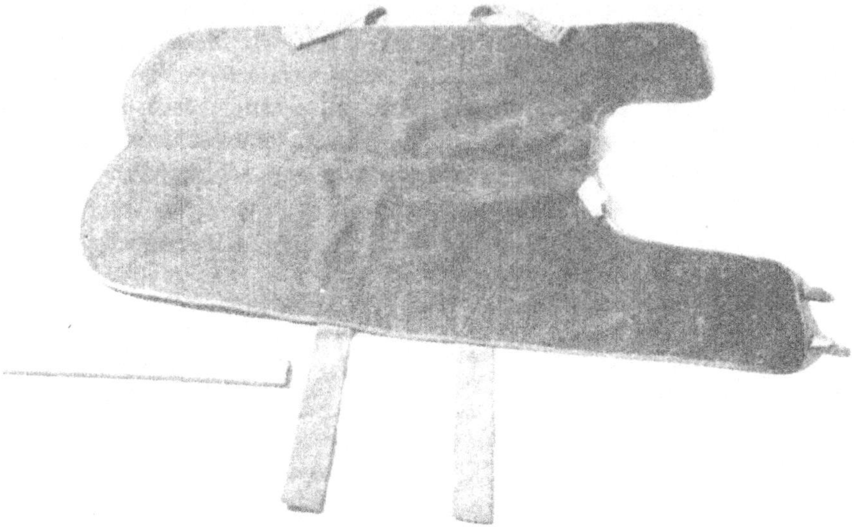

FIGURE 9.—Waterproof blanket—reverse.

One of these periods must be prior to the training period. If the dog is not permitted to perform these functions, he will interrupt the training procedure as well as make the training field unsuitable for work. If he is prevented from evacuating during the training period, and has not previously carried out this function, he is usually unable to give his best training performance. It should be borne in mind that a mature dog cannot empty his bladder thoroughly except by frequent evacuations. He must be allowed time to perform this function thoroughly. Urine is discharged from the bladder at frequent short intervals until pressure is entirely relieved.

SECTION VI

FEEDING

71. Food requirements.—To a great extent, the dog is able to digest and utilize the same foods as man. As far as anyone knows today, the dog has the same nutritional requirements as man, except that man develops scurvy if his diet is deficient in vitamin C, while the dog does not seem to need this vitamin. The dog in the wild state does a very good job of balancing his diet. When he kills a rabbit he devours it entirely; even the head and fur may be eaten. A dog can, however, be maintained in just as good condition on a diet balanced artificially. Proteins contained in the muscles of the rabbit are also to be found in meat, milk products, and soy beans. In place of the plant products found in the stomach of the rabbit, the balanced feed provides carbohydrates through corn or wheat products. Vitamins contained in the liver of the rabbit may also be provided through cod-liver oil or concentrates mixed with the dry feed. Foods may be classified according to what they will supply to the animal's system. In general they are grouped as follows:

a. Proteins.—Flesh-making and tissue-building; mostly supplied to dogs in meat.

b. Fats.—Energy- and heat-producing; supplied in meat fats.

c. Carbohydrates.—Energy-producing; supplied in bread and cereals.

d. Minerals.—Bone-building; supplied in cereals, vegetables, and bones.

e. Vitamins.—Generally necessary; supplied in milk, vegetables, eggs, and meat.

72. Meat.—Meat is considered the essential part of a normal dog's diet. When it is used as a ration, it is almost completely digested and assimilated. Most meat or meat products can be used in the dog's diet; such parts of the carcass which are usually considered unfit for human consumption are, on the whole, more nourishing to the dog than cuts of muscle meat. Intestines, lungs, heart, liver, genital organs, lips, nostrils, cheeks, udders, tail, brain, and tripe are all excellent sources of animal protein essential in the dog's diet. Beef, mutton, lamb, and horse meat are to be preferred; fresh meat is more satisfactory than meat that is salted, cured, or preserved. Horse meat is an excellent food for dogs provided that the meat is not too lean, or that fat is added if there is not enough in the meat. When horse meat is fed, it often causes a relaxed condition of the bowels. This condition may be avoided by gradually adding horse meat to the diet. Meat may be fed raw, boiled, roasted, or broiled, but should not be fried. It may be fed in medium-sized pieces, or ground.

73. Other sources of protein.—Other sources of animal protein are eggs, which may be fed cooked or raw; milk, dried or raw; cheese;

and fish. Fresh fish, particularly cod, salmon, and tuna, makes a satisfactory substitute for meat when given occasionally, provided it does not contain harmful bones. Since the protein content of fish, milk, and cheese is usually somewhat lower than meat, it is necessary to feed larger quantities of these foods to obtain the same protein balance. Meat or other sources of protein must be regularly included in the dog's ration, and should constitute about one-half of the total to insure satisfactory growth of young dogs and repair of tissue in mature animals.

74. Cereals and cereal products.—Cereal grains and certain cereal products are not greatly relished by dogs. Such foods, however, should generally be given, since they supply bulk, energy, protein, some vitamins, and minerals. Cereals may constitute about 25 percent of the total feed of mature dogs. The cereal grains most commonly utilized in some form for dog feeding are corn, rice, oats, wheat, and barley.

a. Corn is used chiefly as meal to make corn bread—a feed that is most suitable for cold-weather rations and for dogs getting abundant exercise. Corn bread is rich in carbohydrates and is not recommended for animals suffering from certain skin disorder.

b. Rice is considered a suitable dog feed by some authorities but not by others. To be satisfactory for use, it must be cooked thoroughly and should be slightly seasoned with salt to increase palatability. Unpolished rice is superior to the common polished type, both in mineral and in vitamin content, but it is also richer in fiber content. Some authorities think cooked rice causes skin troubles similar to eczema, but this may be due to deficiencies in the ration because of an overuse of rice.

c. Oats are sometimes of value for dogs if ground or rolled. Such food must be cooked thoroughly. Like corn bread, it is best for active, outdoor dogs and for cold-weather rations. Injudicious use of oatmeal is said to cause certain skin troubles and intestinal disorders.

d. Except in commercially prepared canned foods, *wheat* is seldom used as a cooked grain for dog feeding, but various products made from this cereal often find their way into the canine ration. Chief of these are bread, some dry prepared breakfast foods, and dog biscuits and meals.

(1) Bread is useful in a variety of forms. Dried or toasted bread that is not needed for human use makes a good cereal food for dogs when given in combination with meat broths, soups, or milk. When fed in this way, only enough liquid should be added to moisten the bread. Bread must never be used if it is moldy.

(2) When prepared breakfast foods made from wheat or other

cereal grains are fed, they are usually moistened with milk, meat broth, soup, or water. Such foods seem particularly suitable for a light morning meal, and they are being used rather extensively.

(3) Dog biscuits of various kinds are also generally suitable for supplying the cereal portion of the ration. They may be fed dry, moistened, or mixed with cooked meat and vegetables. These biscuits can be obtained in many sizes and shapes (square, oval, bone-shaped, cubes, pellets, or kibbled—broken into small pieces), and their composition is variable, depending on the specific use for which they are intended. However, practically all of them are high in cereal and low in moisture content, and they consist of various combinations of meat byproducts, such as legume meals, cod-liver oil, fish meal, molasses, salt, yeast, and other substances. Commercially made dog meals are usually similar to dog biscuits in composition, but they are ordinarily not so well adapted to a variety of uses. They are generally fed mixed with water or milk. They do not by themselves constitute a balanced canine ration, but must have considerable meat added for correct proportions.

e. Barley is not a very palatable dog feed but is used in the cooked form in some commercial mixtures. Barley water and gruel have been found useful for some sick animals and finicky eaters.

75. Vegetables.—The primary functions of vegetables in the dog ration are to furnish vitamins and minerals, supply bulk, act as fillers, and regulate the bowels. Of the common vegetables, tomatoes, canned or raw, are particularly suitable because they can be mashed easily and mixed with various other foods. The extensive use of vegetables having a high fiber content should be generally avoided. Where feeds of animal origin make up one-half of the dog's ration, vegetables may constitute about 25 percent of the remainder, or 12½ percent of the total. If it is necessary to reduce the amount of vegetables, cereals may be used as a substitute for the vegetables removed.

76. Bones.—Bones are absolutely unnecessary to the health of a grown dog. It is true that the average adult dog is able to digest uncooked bones with comparative ease, but the nourishment in a bone is of little consequence when compared with the danger that lies in such a diet. The constant gnawing of bones wears down the enamel-covered crowns of the teeth. If the bones are small or brittle enough to be swallowed, the pieces or splinters may scrape, cut, and even perforate the stomach or intestines. If they pass safely to the rectum, the small particles may pack together and cause an impaction, the result of which may be ulceration, inflammation (chronic proclitis), or prolapse of the rectum.

77. General feeding rules.—The following rules should be generally observed:

a. Number of meals.—Ordinarily, one meal a day is sufficient for a mature dog. If he appears thin and underweight, he may be fed a supplemental ration of milk, eggs or cereal, or all three.

b. Time of feeding.—The time for feeding depends on whether the dog is to work on the night or day shift. He must *never* be fed his main meal just before he goes to work and *never* immediately after strenuous exercise. A dog works best when his stomach is empty. A full digestive tract makes him sluggish.

c. Dryness of meal.—As a general rule, it is desirable to feed dogs on a fairly dry diet. Sloppy food is difficult to digest because the excessive moisture in the food dilutes the gastric juices and retards chemical action on the food. Sloppy food is sometimes vomited by dogs.

78. Feeding procedures in typical situations.—Feeding procedures will vary with the situation in which war dogs are employed. *Wherever possible, the use of good mess scraps is advocated.*

a. Reception and training centers.—When the dog arrives, it is generally desirable, at first, to limit the amount of food and water given him, and then to increase the quantity. This is recommended because major changes in the diet to which a dog has been accustomed are apt to upset his digestion for a few days, causing loose bowels and other symptoms of intestinal disorder. After such adjustment as may be necessary, standard procedure should be followed. The following system has proved satisfactory at some reception and training centers:

(1) *Time of feeding.*—Dogs are fed at 4 p. m. An effort is made to feed as rapidly as possible to relieve the excitement and nervousness of dogs waiting to be fed. All dogs are fed within a 15-minute period.

(2) *Diet.*—(*a*) The formula for a 60-pound dog in training includes—

½ pound cooked horse meat.
¾ pound raw horse meat (ground with the bone).
½ pound yellow cornmeal cooked for 2 hours in horse-meat broth.
½ pound rolled oatmeal cooked for 2 hours in horse-meat broth.
½ pound commercial dog feed.
Salt added in amounts to make 1 percent based on dry weight.

(*b*) As amounts for individual dogs vary, the feeding is supervised carefully by the responsible officer. Individual pans are prepared in the kitchen and transported to the area in wagons. Extra amounts of food accompany the individual pans to take care of the needs of larger dogs and those thought in need of additional amounts, because of thinness.

(3) *Provisions for sick dogs.*—Dogs on sick report are fed specially prepared food containing milk, hamburger, bone meal, eggs, cod-liver oil, and yeast.

b. Field stations.—(1) *Interior posts.*—(*a*) When dogs are stationed at designated posts for interior guard (depots, plants, etc.), they are fed similarly to dogs at reception centers, whenever possible. When exact ingredients of the above formula are not available, the officer or enlisted man in charge of the dogs must prove his resourcefulness by obtaining such substitutes or equivalents for the required foods as is necessary to make up a balanced ration. The following is a sample ration for a dog weighing 50 pounds:

1½ pounds cooked or raw meat, or other sources of animal products equivalent in protein content.
1 tablespoon of mashed vegetables.
2 slices of bread or equivalent amount of a good cereal (either dry or cooked).

(*b*) It is advisable to feed raw meat at least twice a week. In general, a dog's main meal should consist of 1½ pounds of meat to each 50 pounds of body weight, depending upon the size of the dog and the amount of work he is required to do.

(2) *Established tactical posts.*—The same procedure may be followed as for interior posts.

(3) *Tactical posts in the field.*—With tactical units on field service, dogs are issued from ½ to 2 pounds of meat and the balance of the ration is made up from whatever can be supplemented from the mess kit.

79. Water.—Clean, fresh water should be supplied for drinking purposes (fig. 10), although usually a dog is not particular about the source of his water supply and will drink anywhere if he is thirsty. The observing handler will note that a dog will eat the meal provided and immediately go to the water bowl. Usually he will drink a few laps and go away. Occasionally a dog will drink excessive quantities of water and the result is vomiting of meal. In such cases it may be necessary to ration the water in small quantities at frequent intervals. If the vomiting condition persists, it should be brought to the attention of the veterinarian. In watering new arrivals, it is advisable to follow the precautions recommended for their feeding.

FIGURE 10.—Watering.

SECTION VII

KENNELING

80. Types of kennels.—In general, kennels are of two types—

a. Multiple units.—These are to be found as permanent installations at reception and training centers and are used for receiving and isolating dogs (fig. 11). Multiple units are also provided at fixed interior installations.

b. Single unit kennels.—(1) Individual kennels are used with tactical units in the field as well as for dogs in training and at fixed interior installations. Individual kennels usually have a hinged top which may be raised to various heights, ranging from a few inches for ventilation to a fully opened position for cleaning the kennel (fig. 12). They also have a detachable section which serves as a sunning bench, grooming bench, shade, or protection for the kennel entrance in wet weather.

(2) A special type of single unit has been found to be very economical as well as warm and comfortable. The unit consists of a whiskey barrel that has been cleaned, lined with paraffin, and mounted on wooden cradles to keep it on an even keel a short distance above the ground. The cradles are firmly attached to the ground but the barrel rests on them by its own weight, thus permitting it to be lifted off easily and thoroughly cleaned. An entrance is cut into one end of the barrel, which faces the south, and is covered with a blanket, burlap sacking, or other available material. Because of the tight construction of a barrel made to carry liquids, it offers excellent protection against bad weather. A porch with slanting roof may be added for further protection and also to permit the dog to sleep outside comfortably on warm days.

81. Kenneling in the field.—In the field, adequate kenneling depends upon the ingenuity of the individuals in charge. Natural resources, such as palm leaves and fir branches, may be used to construct a crude protection from sunlight as well as rain. In rainy climates the dog's bed should be elevated 6 to 10 inches from the ground (fig. 13). Any tendency on the part of the dog to share the quarters of his master must be discouraged. This practice tends to make the dog familiar with too many men; it is natural for most men to try to

FIGURE 11.—Multiple type kennel.

FIGURE 12.—Single unit type kennel.

FIGURE 13.—Typical kennel for hot, wet climate.

make pets of dogs, and such familiarity is harmful to the discipline that has been developed through the military training of the dog.

82. Kennel care.—Kennels should be thoroughly cleaned once a day. A dog is inherently clean in his habits and he must be encouraged to remain so. A dirty kennel makes a dirty dog. The daily cleaning should be done the first thing in the morning. A second inspection at feeding time is recommended. Before night all droppings will have been removed and, in the case of permanent type kennels, runways carefully hosed. Permanent kennels will be scrubbed thoroughly once a week. Drainage must be watched with care, especially after storms, since the channels fill up and the kennel becomes wet. The ground around individual kennels should be raked, and gravel or loam added as required.

83. Bedding.—Bedding is inspected daily. This bedding, which often consists of straw, is shaken up every day and changed at least once a week. In rainy weather the dog's coat usually wets the straw. Wet or damp straw is removed at once. If a dog is moved from one kennel to another, the old bedding is cleaned out and burned and the kennel disinfected.

84. Alinement of kennels.—Individual kennels must not be placed face to face since this will result in the dogs fretting at one another day and night.

SECTION VIII

PREVENTION OF DISEASE AND FIRST AID OF INJURIES

85. General.—The treatment of disease and injuries among dogs, as well as prevention of disease, is a function of the Veterinary Corps. Many diseases and injuries are preventable if all concerned are vigilant, intelligent, and untiring in the application of simple preventive measures. Frequently the development of serious diseases or injuries can be prevented by prompt first-aid measures and early treatment. In this section, preventive measures and first-aid treatment are briefly discussed.

86. The dog in health.—Health is the condition of the body in which all the functions thereof are performed in a normal manner. It is particularly essential that the normal functions of the body of the animal be thoroughly understood, otherwise one cannot hope to recognize any departure therefrom. Even the most elementary study of disease conditions must be founded upon a very thorough knowledge of the normal body.

a. Appearance.—In the standing position the four feet are placed squarely on the ground or one hind foot may be placed slightly in advance of the other. The head is on the alert, eyes wide open, nose moist and free from discharge, and in prick-eared dogs, the ears are erect. The coat has a luster and the skin is supple and easily moved about over the structure underneath. The color of the mucous membranes of the eyes, nostrils, and mouth is salmon pink. The bowels are evacuated on an average of three to four times in the course of 24 hours and the stool should be formed and free from mucus. The color of the stool varies according to the nature of the food. The urine is voided at frequent intervals and during the course of 24 hours large dogs will secrete 0.5 to 2 liters of urine varying with the food, external temperature, and season of the year. The color will vary from light to dark yellow depending on the diet.

b. Temperature.—The normal body temperature of the dog at rest is 101° F., but may vary 1° in either direction. Immediately after work and especially in a hot sun the temperature of healthy animals may rise, but as a rule it is unaffected by climatic conditions and it is thus a most valuable guide. The temperature is taken with a clinical thermometer in the rectum. The thermometer is moistened or oiled, the mercury is shaken down to 96° F. or below, the bulb of the thermometer is inserted in the anus and allowed to remain for 3 minutes, when it is withdrawn, the temperature read, the thermometer washed in cold water and the mercury shaken down again below normal. A rise of temperature often precedes any visible symptoms in contagious diseases and is, therefore, important in detecting animals suspected of infection.

c. Breathing.—The breathing may be most conveniently noted by the rise and fall of the flanks. The average number of respiratory movements while at rest normally vary from 8 to 16 per minute. These should be counted when the animal is first approached, as excitement or movement cause an increase in the number of respirations. Immediately following exercise, the respiration rate may increase to 60 to 90 per minute. During warm weather, while the dog is at rest, the respirations are greatly accelerated due to the fact that the skin glands are not active enough to assist in the respiratory functions. In order to overcome this condition, the dog breathes through the mouth and the frequency of the respiration is increased.

d. Pulse.—The normal pulse is from 70 to 100 beats per minute, depending on the breed of the dog. The pulse may be felt on the left side just in back of the elbow joint, or by placing the finger tips lightly on the femoral artery which is located on the inside of either hind leg. The beats are counted for 30 seconds and multiplied by 2. The animal should be at perfect rest when the pulse is taken, as exercise and excitement quicken it. In illness the pulse is usually faster than normal and its character varies considerably. It is necessary to note whether the pulse is stronger or weaker than it should be.

e. Mucous membranes of the eyes, nose, and mouth.—Normally they are a salmon-pink color and free from congested areas and discharges, but during illness they may become red (congested), pale or white (from loss of blood), or yellow (from digestive disturbances).

f. Skin.—In health the skin should be supple and roll easily on the underlying structure. The hair should have a luster, feel soft, and lie evenly. In disease the skin is dry as if adhering to the muscles below; the coat is dull in appearance, harsh to the feel, or "staring"; that is, with the hair on end instead of lying flat.

87. Indications of disease.—Every disease has different indications and the symptoms vary so greatly that only exhaustive study can acquaint one with their many indications. In order that sick animals may be detected in the early stages of disease and brought to the attention of a veterinarian, the soldier must be prepared to recognize the more common early variations from normal. The most common preliminary indications of disease are partial or complete loss of appetite, elevation of temperature, accelerated breathing, increased pulse rate, listlessness, dejected appearance, nasal discharge, persistent cought, diarrhea, constipation, unhealthy coat of hair, and unnatural heat or swelling in any part of the body. One of the first and most important symptoms of sickness is impairment of appetite. The best times to inspect animals for evidence of sickness or injury are while they are being fed or groomed.

88. Classes of diseases.—*a. Communicable diseases.*—(1) *Nature.*—Communicable diseases are diseases that are transmitted from animal to animal either by direct contact between the sick and well, or indirectly through the medium of infected kennels, unsanitary feeding dishes, flies, lice, ticks, mosquitoes, and faulty sanitation. The diseased animals throw off in the discharges from the respiratory, digestive, and urinary systems the material which will cause disease in susceptible animals. These diseases deserve more attention than noncommunicable diseases among war dogs because they are most likely to appear when dogs are congregated in considerable numbers. Some of these diseases are incurable, others may be transmitted to man, and all may cause great harm if not held in check. Communicable diseases are always marked by a period of incubation, which is the time interval that elapses between infection and the appearance of the symptoms of the disease. This period may vary from a few days to several weeks. Examples of contagious diseases are distemper, rabies, mange, and ringworm.

(2) *Predisposing causes.*—Certain causes or conditions that lower the vitality and natural resistance of dogs to disease, thereby rendering them more susceptible to infection, are termed predisposing causes. The principal causes are exposure, general debility, improper grooming, lack of sufficient food, overwork, and other diseases.

(3) *Prevention.*—The logical way to prevent the entrance of a communicable disease is to correct the faulty conditions that predispose the animal. When a disease once gains entrance to a group of dogs there are certain rules of procedure that have been found necessary in checking the spread to healthy animals and in stamping out the disease. These measures are as follows:

(*a*) Daily inspection of all animals in order to detect new cases. This insures prompt removal of the sick as a source of infection and the initiation of proper treatment.

(*b*) Quarantine of exposed animals.

(*c*) Isolation of sick animals.

(*d*) Disinfection of infected kennels and the surrounding area, equipment, and utensils.

(4) *Quarantine.*—(*a*) Quarantine is the separation of the apparently healthy dogs that have been exposed to the infection from those that are healthy and have not been so exposed.

(*b*) Animals that have been exposed or suspected of exposure to a communicable disease are potential sources of infection and they may be in the incubative stage of the disease. In many diseases the dog may be infected during this incubation period or in the very early stages of the disease before the characteristic symptoms are noticeable. For this reason such animals should be placed in quarantine in order to protect the nonexposed animals from this possible source of infection.

(*c*) In most of the ordinary communicable diseases of dogs the period of incubation is less than 3 weeks. It is obvious that the period of quarantine should be longer than the period of incubation.

(*d*) A uniform quarantine period of 21 days has been adopted by the Army. The discovery of a new case in the quarantine group is cause for beginning a new 21-day period of quarantine. The quarantine of exposed animals during an outbreak of disease is mandatory, necessary and prescribed in AR 40–2090. Provisions for the quarantine of newly arrived animals at a station are prescribed for in AR 40–2035.

(*e*) The place selected for quarantine purposes should be located so that it is impossible for other dogs to enter or come in contact with dogs in such area.

(*f*) The severity of the quarantine rules depends upon the nature of the disease. Attendants in the quarantine area should not handle other dogs, and strict sanitation is mandatory at all times.

(5) *Isolation.*—Isolation is the absolute segregation from all other animals of an animal affected with a communicable disease, or one suspected of being infected. It must be complete in every detail in order to be of any value.

(6) *Disinfection.*—The application of agents called disinfectants, used for the purpose of destroying disease producing organisms, is called disinfection. It is essential that a thorough cleaning always precede disinfection. It is essentialy time wasted to disinfect a kennel, utensils, or anything used by a dog unless it has previously been thoroughly cleaned. Fresh air and sunlight are potent disinfectants. Thus a kennel with plenty of windows and ventilation is more sanitary than one that is dark and damp.

b. Noncommunicable diseases.—Noncommunicable diseases include all diseases that are not transmissible either directly or indirectly from one animal to another. Many of these diseases are directly caused by improper methods of animal management. While noncommunicable diseases are the cause of many lost days of animal service and animal losses, they do not present the serious problems encountered in communicable diseases. Examples of noncommunicable diseases are nephritis, epilepsy, and nutritional and digestive disturbances.

89. Nursing.—*a.* Good nursing is indispensible in the treatment of sick and injured animals. The chief points to consider in nursing are:

(1) *Ventilation.*—Allow plenty of fresh air but protect the animal from drafts. Avoid exposure to extreme temperature and in the field provide shelter from wind and rain.

(2) *Bedding.*—A good clean bed induces an animal to rest. It should be changed several times daily and all food particles and fecal material removed.

b. Convalescent patients should receive just as much exercise as each individual case permits. It must be borne in mind, however, that absolute rest is one of the best treatments.

c. Dogs that are weak and depressed should not be worried with unnecessary grooming. Such animals should be carefully brushed once a day and the eyes and nostrils wiped out with a moist piece of cotton. Animals that are only slightly indisposed should be groomed.

d. Some sick animals retain a good appetite. The principal things to observe in such patients are that they are not overfed, that the stools are kept soft, and that they have plenty of fresh water. Sick animals with impaired appetites require special attention and often relish a change in diet. The feeding dishes should be kept clean. Feeding should be often and in small amounts. Uneaten portions should not be allowed to remain with the dog in the kennel. To induce eating, hand-feeding or forced feeding of liquid foods may be used as a last resort.

90. Wounds and injuries.—*a. Wounds.*—A break in the skin, body tissues, or lining of a body cavity, resulting from external violence or muscular activity of the body itself, is known as a wound. For the purpose of description, wounds are classified as incised wounds or cuts, lacerated wounds or tears, punctured wounds or holes, and abrasions.

(1) Incised wounds or cuts are caused by sharp objects. Although this type of wound bleeds freely due to the severance of blood vessels, there is very little tissue destruction and infection occurs infrequently.

(2) Lacerated wounds or tears evidence tissue destruction and infection. Wounds of this type are caused by falls, striking blunt objects, and dog fights. Bleeding is not as severe as in incised wounds.

(3) Punctured wounds or holes may be caused by any object that will penetrate the tissues. Nails, sharp pieces of wood, or bullets are some of the common causes. Wounds of this type are difficult to clean out and favor infection due to extensive tissue destruction and foreign material which is carried into the deeper tissues. Hemorrhage in this type of wound is of little concern unless a large blood vessel is severed.

(4) Abrasions are wounds caused by the rubbing of the skin against some object which produces an area of irritation. Rope burns are an example.

b. First-aid treatment of wounds.—(1) If bleeding is present, it must be controlled before any attempt to clean or bandage the wound is made.

(2) Bleeding is controlled by the application of a tourniquet or pressure directly to the bleeding surface. A tournquet may be ap-

plied to the legs or the tail of a dog to control hemorrhage. It may be fashioned from a strip of gauze bandage, handkerchief, or necktie and is placed on the side of the wound nearer the body. Loosen the tourniquet at 15-minute intervals to relieve pressure, but do not remove it. If bleeding has been controlled, proceed to clean and dress the wound. Should the bleeding continue, tighten the tourniquet again. When it is not possible to apply a tourniquet, bleeding may be controlled by pressure applied directly to the bleeding surface. Do this by covering the bleeding surface with a clean piece of gauze or handkerchief and placing a compress over the wound, binding it firmly in position with a bandage.

(3) When bleeding has been controlled, proceed to clean the wound. The hair is clipped with scissors and the dirt washed away from the injured parts with clean water. All visible foreign material should be removed from the wound but the wound should not be packed, probed, sutured, or covered with ointment. If available, an antiseptic may be applied and the wound covered first with gauze, then cotton, and then bandaged. Never place cotton directly over a wound, and never apply a bandage over a tourniquet. When first-aid measures have been administered, transport the dog to the veterinary hospital for further examination and treatment.

c. Injuries.—Injuries are classified as bruises, strains, sprains, and fractures. They are caused by external violence or from muscular activity of the body itself, and involve muscles, tendons, ligaments, joints, organs, and bones of the body. A break in the continuity of the tissue covering the body need not accompany an injury.

d. First-aid treatment of injuries.—The application of hot or cold fomentations to bruises, sprains, and strains will prove beneficial. Complete rest is essential.

(1) *Fractures.*—The larger number of fractures take place in the extremities of the body. The causes are varied, but the greater number of cases result from falls, kicks, bites from other animals, and gunshot wounds. Before attempting to administer first aid, secure the dog's mouth with a muzzle. Assuming that the dog has sustained a leg injury, place the dog in a reclining position and carefully examine the leg to determine the extent of the injury. When the leg is found to be fractured (broken), or a fracture is suspected, obtain two flat pieces of wood and fashion splints the length and shape of the leg. Place one splint on the inside and the other on the outside of the leg and secure them in position with strips of gauze bandage. The dog should then be transported to the veterinary hospital. When the fractured area will not lend itself to splinting, place the dog on a firm improvised litter made from boards and transport the dog to the veterinary hospital.

(2) *Eye injuries.*—(*a*) Injuries to the eye are very common. They occur as wounds, lacerations, contusions, and result from contact with foreign objects or external violence such as bites received in dog fights.

(*b*) In rendering treatment for an eye injury, the eye should be bathed with clear warm water and then examined for the presence of any foreign material. When present, it should be carefully removed provided such action will not cause further injury. A moist piece of gauze should then be placed next to the eye and cotton and bandage applied.

(*c*) Occasionally a dog's eye will be prolapsed or forced outside the lids. Dogs with prominent eyes are predisposed to this condition which may result from injury or fighting. An attempt should be made to replace the eye immediately. This is done by grasping the lower lid firmly with the thumb and fingers and pulling outward. At the same time place a piece of clean gauze which has been moistened with mineral oil on the eyeball and push steadily and firmly inward. Usually the eyeball will slip back into position. If this cannot be achieved within a few minutes, moisten the eyeball thoroughly with mineral oil, cover with gauze, cotton, apply a bandage and transport the dog to a veterinarian.

(3) *Injuries due to burns.*—(*a*) Burns may be caused by hot liquid, chemicals, fire, and friction. The extent of the damage to the body tissues caused by burning is described by the appearance of the injured parts. In first-degree burns the skin is reddened; blistering occurs with second-degree burns; when deeper destruction occurs, the tissue takes on a cooked appearance and the condition is referred to as a third-degree burn.

(*b*) As soon as first-aid measures have been completed, immediate treatment is required as death will often occure within 1 or 2 days from shock.

(*c*) In applying first-aid treatment, procure several clean pieces of cloth or several layers of sterile gauze. The cloth or gauze is soaked in a solution of Epsom salts or bicarbonate of soda and applied over the injured area. These solutions are made by adding two tablespoonfuls of Epsom salts or bicarbonate of soda to a pint of clean, warm water. Bandage the gauze lightly in place and cover the dog with blankets to keep him warm until further examination and treatment is rendered by the veterinarian.

91. Disturbances of digestive system.—*a. Foreign bodies in mouth.*—When dogs claw at the mouth, have difficulty eating, or have an excessive flow of saliva from the mouth, the presence of foreign objects should be suspected. Pieces of bones, splinters of wood, and small stones will lodge between the teeth and cause the dog distress. Caution must be exercised when dogs evidence symptoms of this nature as similar symptoms are noted in dogs having rabies. Cases of this kind should be diagnosed by the veterinarian.

b. Disturbances of digestive tract.—Digestive disturbances are characterized by loss of appetite, persistent vomiting, diarrhea, stools streaked with blood or covered with mucous, and constipation. These symptoms may be due to faulty diet, ingestion of poisonous substances, infections, foreign bodies in the intestines, and parasitic infestations.

(1) *Persistent vomiting and diarrhea.*—When these conditions exist, food should be withheld and the quantity of drinking water limited. Purgatives should not be administered until the cause has been determined.

(2) *Constipation.*—It is usually the result of heavy feeding, insufficient exercise, and the use of bones and dry biscuits in the diet. To encourage an elimination, a suppository fashioned from a piece of noncaustic soap may be inserted into the rectum. A soft diet should be fed and bones and biscuits eliminated. Two tablespoonfuls of mineral oil administered morning and night for 2 or 3 days will prove beneficial.

(3) *Poisoning.*—When it is known or suspected that a dog has eaten some poisonous product, vomiting should be induced immediately. Accomplish this by administering large quantities of soap water, salt water, or dish water. When the dog has vomited freely, administer a tablespoonful of Epsom salts diluted in water. If a poison such as bichloride of mercury has been eaten, follow the above treatment and then administer large quantities of milk. Keep the dog warm until further treatment is administered by a veterinary officer.

(4) *Intestinal parasites.*—Heavy infestations of intestinal parasites will cause digestive disturbances and loss of condition. Those commonly found in the digestive tract of a dog are round worms, tape worms, hook worms, and whip worms. Occasionally the presence of parasites may be detected upon the physical examination of the dog's stool, and frequently tape-worm segments will be noticed adhering to the hair around the anus. "Shot gun" worm remedies should not be administered to a dog suspected of or known to have intestinal parasites. Vermifuges are toxic preparations and may produce drastic reactions or fatalities when the dogs have not been properly prepared for worming or intestinal irritation is present. A microscopic examination of the feces by the veterinarian will reveal the presence of parasitic ova and specific treatment will be prescribed.

92. Disturbances of nervous system.—The causes of the various nervous system derangements common to dogs, such as hysteria and convulsions, vary from digestive disturbances, excitement, and malnutrition, to the organisms causing infectious diseases. When a dog evidences symptoms of hysteria, convulsions, or general nervous disturbances, do not attempt to apply restraint or administer treatment. If the dog is on a leash, secure him to a stake in a shaded area or place him in a kennel and allow the condition to subside. Summon the aid of the veterinary officer or, if the dog's condition returns to normal, transport him to the veterinary hospital for observation.

93. External parasites.—Fleas, lice, and ticks are parasites which are found on the body of the dog.

a. Fleas.—(1) Fleas reside in insanitary surroundings. They attach themselves to dogs and receive their nourishment by blood sucking. They cause the dog considerable annoyance and in large numbers may produce chronic skin eruptions. They also cause the dog to inflict self-injury through the medium of persistent scratching.

(2) The flea is intermediate host of the dog tapeworm. When dogs ingest fleas harboring tapeworm eggs, the flea is digested, the tapeworm eggs develop into larvae, then adult parasites form in the digestive tract of the dog.

(3) Thorough grooming, strict sanitation, and disinfection are required for the elimination of fleas. The destruction of fleas on the dog may be effected by the application of flea powder, or by bathing of the dog in a 1- to 2-percent solution of creolin.

b. Lice.—Two varieties of lice have been found to infest the dog—biting and sucking. When dogs are found to be infested with lice they should be isolated. The dog should be clipped and the hair destroyed by burning. All dogs in the kennel should be thoroughly groomed every day to detect further evidence of this condition. The kennels should be thoroughly disinfected and all bedding destroyed by burning. Cleanliness discourages lice propagation. In the treatment of long-haired dogs affected with lice they should be clipped and bathed in a 2-percent solution of creolin or other suitable parasiticide solutions.

c. Ticks—(1) The common dog tick is a blood-sucking parasite which attaches itself to the dog by burying its head in the skin. To facilitate the removal of ticks, moisten a piece of cotton with ether, place it over the tick and draw gently but firmly away from the dog's body. Jerking the tick away from the dog's body will allow the head of the tick to remain imbedded in the tissues and produce further irritation. The area from which the tick is removed should be washed and an antiseptic applied.

(2) The female tick is gray and the male is brown in color. Frequently they will be found attached to one another or feeding on the same area. When ticks are removed from the dog, place them in a receptacle and upon completion of grooming they should be destroyed by burning.

(3) Ticks are carriers of disease transmissible to man and care should be exercised in handling this parasite.

(4) Careful grooming and removal of all ticks from all dogs should be carried out daily. Enforcement of strict sanitary measures and the daily application of parasiticides to infected kennels are required.

94. Bandaging.—*a. Uses.*—Bandages are one of the oldest forms of surgical apparatus and are employed under varying conditions for many purposes. The chief uses for bandages are to—

(1) Control bleeding by pressure.

(2) Hold dressings and compresses in place.

(3) Hold splints in place.

(4) Immobilize a part of the body.

b. Application.—(1) The roller bandage is the type most frequently used for applying dressings, splints, compresses, and pressure.

(2) In applying a bandage, use one hand to hold and direct the roll, the other to keep each lap smooth and to hold the turn while taking up slack or changing directions. Apply the bandage to the injured part snugly enough to exert the desired amount of pressure and to prevent slipping, but be careful not to shut off circulation by too much pressure. When the ends of the bandage are secured they should be tied only as tight as the bandage has been applied.

(3) A bandage should be applied over gauze and cotton and not directly to the body. It is not advisable to apply a wet gauze bandage, for in drying it may tighten and cause undue pressure. Bruised areas usually become swollen within a short time following injury. This should be borne in mind when bandaging is contemplated, as the application of a snug bandage to the original wound will impair circulation if swelling occurs.

(4) In applying a bandage to the leg it is advisable not to bandage the foot unless it is involved in the injury. Tight bandages will cause swelling in the foot and the condition can quickly be recognized and corrected if the foot is exposed. In securing the ends of a bandage, fasten them on the outside of the leg but not directly over the injured area, as pressure from the knot on the wound will cause the animal discomfort.

(5) When it is desired to apply a bandage to the body itself, a many-tailed bandage is effective. This is made from a piece of cloth long enough to completely encircle the body and permit the making

of ties. The material is folded in half and strips cut into the material at the open ends. The dressing is placed on the chest, flank, or abdomen, the bandage placed around the body, and the strips tied on top of the back. Made on a smaller scale, a many-tailed bandage may be applied to the head or neck.

95. Reporting dog bites.—*a.* Due to the prevalance of rabies in certain communities, it is important that all cases of dog bites be reported immediately.

b. Rabies is of particular importance because it is readily transmissible to man through bite wounds inflicted by animals, particularly by dogs.

c. Rabies is spread from animal to animal through the medium of infected dogs which run at large and attack healthy dogs. Dogs bitten by strays, or by other Government-owned dogs should be placed in the custody of the veterinarian and quarantined as prescribed in AR 40–2090.

96. Restraint of dogs.—*a.* Before administering first aid to the injured dog it is always advisable to place a muzzle around the dog's mouth. No matter how well the soldier is acquainted with the normal behavior of the dog, it must be remembered that an injured or frightened dog will often bite through intuition as an act of self-defense.

b. The muzzle is fashioned from a piece of gauze bandage or from a firm strip of cloth. It is placed around the mouth and looped under the lower jaw. The ends of the bandage are pulled taut to tighten the loop and then passed to the back of the neck just below the ears, where they are securely tied.

CHAPTER 3

BASIC TRAINING

SECTION I

PRINCIPLES OF DOG TRAINING

97. General.—There are no tricks or mysteries to dog training. It is a relatively simple process if based on—

a. A practical knowledge of how a dog's mind works.

b. Constant repetition of training exercises.

c. Suitable recognition of a dog's progress.

d. Patience.

98. Basic principles.—The effectiveness of a dog training program depends on the regard shown for certain basic principles.

a. The trainer must establish himself as the master of the dog or dogs assigned to him. He pets, praises, feeds, and handles only the dog or dogs assigned to him; he does not permit any individual other than himself to make friends with the dog or dogs assigned to him.

b. The trainer must be aware of the limitations of a dog's mind. There are many things one cannot expect of a dog and many things one can, provided he is properly handled. In the beginning, a dog may be uncertain of what is expected of him. That is normal. A command may not be fully "understood" until the dog has been made to carry it out numerous times.

c. There are specific techniques for giving commands so that they are suggestive or meaningful to a dog.

(1) *Vocal commands.*—Vocal commands are given firmly and clearly. Tone and sound of voice, not volume, are the qualities that will influence the dog. They must be directed at the one dog concerned—not voiced in any direction. They must inspire obedience.

(2) *Gestures.*—Next to the voice, gestures are the chief means of influencing dogs. Often they are combined. When training is first undertaken gestures may be exaggerated to help convey the desired command to the dog. As training progresses exaggeration is reduced.

d. It is essential that the dog be made to carry out the same command over and over until he can make the desired response without delay. Repetition is even more important in dog training than it is in human learning. However, both trainer and dog can go stale or lose efficiency by practicing any one command too much during one period. It is better to go on to another exercise or let some time elapse before returning to practice the command in question.

e. The trainer must never lose patience or become irritated. If he does, the dog will become hard to handle because he takes his cue from the trainer. Patience is one of the prime requisites of a good dog trainer, but it must be coupled with firmness. The moment the dog understands, obedience must be demanded if the dog is to be a prompt and accurate worker. Dog training takes time and understanding.

f. From the very beginning of training, the dog should never be permitted to ignore a command or fail to carry it out completely. He must learn to associate the trainer's command with his execution of it. He should never be allowed to suspect that there is anything for him to do *but* obey. He must learn that he will have to do what the trainer commands, that he will have to carry out the command completely, no matter how long it takes. Laxity on the part of the trainer on even one occasion may result in an attitude or mood of disobedience that means difficulty and delay in the continuation of the training program. When the trainer is sure that a dog knows what is expected of him and is being willfully stubborn, the trainer may handle the leash firmly and, if necessary, punish the dog.

g. The aim of punishment is improvement, not reprisal.—(1) A dog does not understand abstract principles of right and wrong according to human standards, and reward and punishment are the means of teaching him what it is intended that he learn.

(2) It is seldom necessary to resort to physical punishment to teach a lesson to a sensitive dog. Withholding of praise, a rebuking tone, or even "No" said reprovingly, are usually sufficient punishment for him. If the dog is callous or insensitive, punishment in his case must be more severe. The punishment must be made to fit the dog as well as the misdeed. Timing in punishment is very important. The correction, whatever form it takes, must always be administered immediately the dog misbehaves. A dog cannot connect punishment with a misdeed committed at some time previous to the punishment.

(3) Real punishment should be inflicted as a last resort and only for deliberate disobedience, stubbornness, or defiance when the dog has learned better. He must never be punished for clumsiness, slowness in learning, or inability to understand what is expected of him. Punishment for such reasons, instead of speeding training, will have the opposite effect. The word "No" is used to indicate to the dog that he is doing wrong. "No" is the only word used as a negative command. It is spoken in a stern and reproving tone. If this form of reproof is not successful, the dog should be chained or kenneled. A dog never is slapped with the hand or struck with the leash. The hand is an instrument of praise and pleasure to the dog and he must never be allowed to fear it; beating with the leash will make him shy of it and lessen the effectiveness of its legitimate use. The dog's name is never used in connection with a correction.

h. Whenever a dog successfully executes a command, even though his performance has taken more time than desirable, the trainer always rewards him with a pat on the head and praises him in some obvious way.

(1) Dogs are usually anxious to please. They must be shown how to do so. When a dog is rewarded for his performance he senses that he has done the right thing and will do it more readily the next time he is given the same command. Praise may have the following forms:

(*a*) Kind words.

(*b*) Patting.

(*c*) Allowing a few minutes romping.

(*d*) Allowing dog to perform his favorite exercise, including free run and play.

(2) It is generally inadvisable to reward a military dog by feeding him tidbits, as he will become accustomed to this form of reward and expect it for some act performed in the field where such food is not available.

(3) Every training period must conclude with petting, praise, and encouragement for the dog, in order to keep up his enthusiasm for his work. If the dog's performance of the actual exercise does not warrant this, he must be allowed to perform a short exercise which he knows thoroughly and does well, so that he will earn reward legitimately.

SECTION II

QUALIFICATIONS OF TRAINING PERSONNEL

99. Essential traits.—Successful care and training of dogs depends to a great extent on personal characteristics of the trainers. Experience has shown that the following traits are essential:

a. Friendly attitude toward dogs.—Any individual selected for the training of dogs should be sympathetic and friendly toward dogs. This is a primary requisite.

b. Intelligence.—It has been demonstrated that individuals with less than average intelligence cannot be taught to care for and train dogs successfully.

c. Patience and perseverance.—The trainer cannot force desired behavior upon dogs nor can he expect dogs to learn as readily as human beings. He must therefore be patient, and he must persevere until each exercise is brought to a successful conclusion.

d. Mental and physical coordination.—A good trainer must be able to convey his wishes to the dog by body movement and gestures as well as by voice. This requires a definite amount of mental and physical coordination.

e. Physical endurance.—Not only must the trainer be able to show good coordination, he must also be able to maintain his efforts as long as necessary. The trainer must "outlast" his dog during each training period.

f. Resourcefulness.—Although training procedure has been carefully set forth in this manual, it is inevitable that situations will arise calling for action not covered by rules.

g. Dependability.—The welfare of the dog is entirely in the hands of the trainer or master. Dogs cannot tell how they are being treated nor can they make reports. Their physical well-being depends, furthermore, on the willingness of the trainer or master to do such manual labor as is necessary for kennel management, feeding, and dog cleanliness. Failure in these responsibilities means failure of the training program.

100. Determination of qualifications.—There are no purely objective methods of determining the extent to which a prospective trainer possesses each of the foregoing traits, yet prospective trainers cannot be selected haphazardly. Interviews provide a satisfactory basis for selection if carefully conducted by a responsible and duly qualified officer or civilian trainer. Each candidate should be examined with regard to all of the aforementioned essential traits before he is permitted to undergo any training. To insure the effectiveness of this procedure, each candidate should undergo a second interview after the first 5 days of his training. At this time, the candidate should be able to give a reasonably clear and intelligent account of the instruction he has received and his attitude toward dog training should be carefully checked. If the candidate is rated "Unsatisfactory" upon conclusion of this interview, he should receive no further training and should be returned to his proper station.

Section III

OUTLINE OF BASIC TRAINING PROGRAM

101. Objectives.—The objectives of basic training are to—

a. Develop in dogs behavior that is basic to more specialized training for specific military functions.

b. Determine the specific military function for which each dog should be trained.

c. Simultaneously teach specially selected enlisted men to—

(1) Train dogs.

(2) Train other men in the using agencies to which the dogs may be assigned.

102. Scope.—The basic training program involves—

a. Training dogs to carry out the following fundamental commands:

(1) Training on leash—

(*a*) HEEL.

(*b*) SIT.

(*c*) DOWN.

(*d*) COVER.

(*e*) STAY.

(*f*) COME.

(*g*) CRAWL.

(*h*) JUMP.

(2) Training off leash—

(*a*) DROP.

(*b*) JUMP.

b. Accustoming dogs to—

(1) Muzzles and gas masks.

(2) Riding in a car.

(3) Gunfire.

103. Personnel.—Three classes of personnel are involved in the training of dogs.

a. Instructors.—Qualified experts, either officers or civilians, designated to take charge of training men and dogs in each branch of specialized training.

b. Assistant instructors.—Individuals designated by each instructor to assist him in training his classes.

c. Trainers.—Men undergoing training in classes conducted by instructors and assistant instructors. These men simultaneously train dogs.

104. Equipment.—*a*. The chief equipment in the basic training program consists of the following:

(1) *Leather leash.*—This should be about 6 feet long from the end of the snap to the end of the hand loop, with a strong snap. The hand loop should be about 6 inches long (fig. 14).

(2) 25-foot leather, rope, or flat-webbing long line (fig. 15).

(3) *Chain choke collar, with strong welded rings.*—It must be large enough to permit easy insertion of the trainer's hand between the chain and the dog's neck. The chain choke is the regular working collar (fig. 16).

(4) *Chainette, or throwing chain.*—This is a plain chain, closed at both ends, and devoid of any sharp edges. It is used for correcting a dog at a distance from the trainer (fig. 17).

(5) Movable hurdle, 3½ feet high and 4 feet wide. The uprights are slotted so as to hold 6-inch boards which are removable to diminish the height of the hurdle as necessary. The two bottom boards may be permanently attached.

(6) *Muzzle.*—This should be available if needed for an individual dog (fig. 18).

(7) *Canine gas mask.*—This should be available for training purposes.

b. Proper use of this equipment is of the utmost importance since it is by these means, combined with voice and gestures, that the trainer communicates his wishes to the dog and controls and disciplines him.

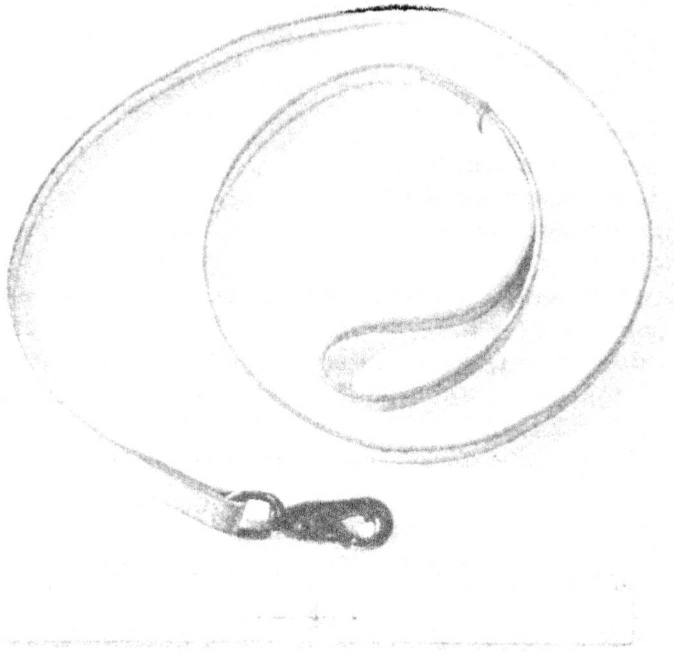

FIGURE 14.—Standard leather leash.

FIGURE 15.—25-foot long leash.

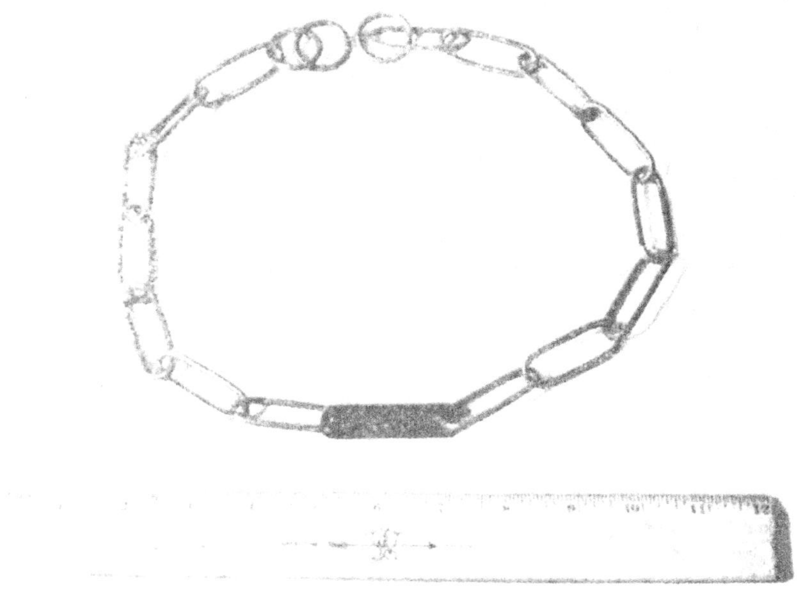

FIGURE 16.—Choke chain leash.

FIGURE 17.—Chainette, or throwing chain.

105. Facilities.—Training classes must be held within an inclosed area.

106. General procedure.—*a*. As soon as a dog is assigned to a trainer, he takes his dog for a walk on a leash. This is done to help establish mutual understanding before training starts.

b. Instruction is organized as follows:

(1) The working unit for basic training is the squad, with the assistant instructor of each squad serving as corporal.

(2) At the beginning of a training period each assistant instructor takes his squad of trainers to the kennels for the dogs they are to work with for that period.

(3) Under the supervision of the assistant instructor, the trainers take their dogs on leash and fall in line, with each dog on the left side of the trainer and an interval of 6 feet between trainers.

FIGURE 18.—Muzzle.

(4) From this point training proceeds, as outlined in this manual, under the direction of the instructor or the assistant instructor, or both.

(5) At the end of the training period, the assistant instructor leads his squad back to the kennels where another group of dogs are assembled for the next period.

(6) Each training period lasts about ½ hour. Ordinarily, each dog is given two training periods daily—one in the morning and one in the afternoon.

(7) Every basic training period starts with a short exercise of commands which the dog has already been taught. In this way the dog's mastery of the basic responses is strengthened. Furthermore,

starting the training period with familiar activities facilitates subsequent learning, by instilling confidence and obedience in the dogs. These review periods can be shortened to about 5 minutes as training advances.

Section IV

HEEL (ON LEASH)

107. Standard of performance.—*Heeling.*—At the command Heel, the dog of his own volition walks at the trainer's left side with

Figure 19.—Executing the command Heel (on leash).

his right shoulder even with the man's left knee. The leash hangs loose. The dog must heel correctly whenever the trainer changes his direction or gait, or when he walks through a group of people (fig. 19).

108. Training procedure.—*a. Initial exercises in heeling.*—After the squad has fallen in, the instructor or assistant instructor gives the command Forward—hup, whereupon each trainer gives his dog the command Heel in a firm voice, jerks slightly on the leash, and steps off

FIGURE 20.—Group exercise in heeling, at the command TO THE REAR, MARCH.

smartly, keeping 6 feet distant from the trainer on his right. The group walks around a prearranged training square, the trainer at the head of the column changing direction at the corners without command. The group walks around this square several times. During this period each trainer repeats the command HEEL to his dog every two or three steps, and endeavors continuously to get the dog to walk easily and on a loose leash at his side. If the dog gets ahead of his trainer, or falls behind him, the trainer jerks on the leash and repeats the command HEEL. He never pulls steadily on the leash, nor allows the dog to pull on it, but rather jerks the leash repeatedly when necessary, allowing it to fall slack immediately after each jerk. A steady pull upon the leash merely causes the dog to set his muscles against the collar and makes the taut leash ineffective for training purposes. If the dog pulls ahead, causing the leash to become taut, the trainer drops his hand to lessen the tension, and then jerks the leash.

b. Advanced exercises in heeling.—To vary the monotony of practice in heeling, as well as to teach the dogs to stick close to their trainers, "left face," "right face," and "to the rear" exercises are used.

(1) *Left face and right face.*—At the command LEFT (RIGHT) FACE—HUP, each man makes a quarter turn in the desired direction, and gives the command: HEEL.

(2) *To the rear.*—As the group is heeling in single file around the training square, the instructor or assistant instructor gives the command: TO THE REAR—HUP. Each trainer then takes one more step forward and does a right-about. As he turns, he passes the leash to his right hand, gives the dog the command: HEEL, and jerks on the leash as he passes it behind his body to his left hand. This causes the dog to follow the leash around the trainer and come to heel on his left side (fig. 20).

c. Amount of training.—Practice in heeling is continued in periods of about 15 minutes until perfection is achieved.

SECTION V

SIT (ON LEASH)

109. Standard of performance.—At the command SIT, which is given when the dog is either standing or lying down, the dog must promptly assume a sitting position. The command may also be given when the dog is heeling (fig. 21).

110. Training procedure.—*a. Initial exercises in sitting.*—After several turns around the square, the instructor or assistant instructor gives the command : HALT. As the trainer stops, he holds the leash in his right hand, leaving his left hand free to ease the dog into a

FIGURE 21.—Correct position for SIT (on leash).

sitting position. He then pushes the dog's hindquarters downward with his left hand and pulls upward on the leash with his right hand (fig. 22).

b. If the dog does not sit facing directly forward, turns either to the left or to the right, or sits facing the trainer, he swings the dog's body around with his left hand in order to get him into the proper position. If the dog sits behind the trainer or too far from his side, he pulls the dog's head close to his left side by using the leash; his left hand restrains the dog from getting up and following the leash. This corrective measure is most effective when carried out as the dog is in the act of sitting, not after his hindquarters have actually touched

FIGURE 22.—Assisting the dog to SIT (on command).

the ground. Each trainer must insist on the proper position at every halt; the instructor or assistant instructor should allow sufficient time for this before giving the order to march again (FORWARD—HUP). When this order is given, the trainer gives the command: HEEL, as before, and moves out. This exercise is repeated as often as necessary to make automatic the response to the command SIT.

Section VI

DOWN (ON LEASH)

111. Standard of performance.—At the command Down, which is accompanied by a downward motion of the arm, the dog must lie

Figure 23.—Executing the Down command from the "sit" position. (Initial stage.)

down promptly. The dog is expected to make this response whether he is standing, sitting, or heeling (figs. 23 and 24).

112. Training procedure.—*a. Formation.*—The command Down can best be taught with the trainers lined up in front of their dogs. Each man issues the command Down to his dog, while the instructor or assistant instructor goes up and down the line helping and criticising.

b. Method.—As the trainer says "Down," he grips the dog's leash close to the collar and jerks down on it. If the dog braces himself and refuses to go down, the man pulls his forelegs out with his free hand and pulls down on the leash at the same time repeating "Down."

As soon as the dog is down the trainer praises and pats him. He then takes the end of the leash in his hand, stands before his dog, and commands: SIT. If the dog does not sit, the trainer pulls upward on the leash, steps closer to the dog, and repeats the command SIT. As soon as the command has been properly executed, he praises the dog again.

113. Amount of training.—This exercise should be repeated 10 or 12 times in a period. After the first period, series of exercises

FIGURE 24.—Executing the DOWN command from the "sit" position. (Final stage.)

devoted to this command, as well as HEEL and SIT command, can be alternated or practiced concurrently. DOWN is a depressing exercise and must not be repeated too often in succession.

SECTION VII

COVER (ON LEASH)

114. Standard of performance.—At the command COVER from the instructor or assistant instructor, the trainer commands: DOWN and as the dog drops, the trainer also drops to the ground at the side of the dog (fig. 25). In tactical situations the dog's master will frequently have to take cover suddenly alongside his dog and the dog must not be alarmed or confused by his action.

115. Training procedure.—If Down has been well taught, there will be no difficulty with Cover. When the exercise is first attempted the trainer does not drop too suddenly lest the dog become frightened

Figure 25.—Executing the command Cover.

and attempt to jump up and run away. After two or three trials, the dog will not be alarmed.

Section VIII

STAY (ON LEASH)

116. Standard of performance.—*Stay.*—At the command Stay, the dog stays in the same position held when the command was given, while the trainer continues ahead. The dog must remain in this position until the man returns to him and gives a different command. The command Stay may be given while the dog is standing, sitting, or down.

117. Training procedure.—*a. Stay (down).*—(1) Practice in staying should be started in the down position since this is the easiest one for the dog to maintain (fig. 26). Each trainer gives his dog the command: DOWN. When the dog has obeyed, the man backs away slowly. As he does so he watches the dog attentively, points at him with his finger, and repeats in a stern voice the command STAY, drawing out the word to cover his retreat. The trainer reaches the end of the leash, always keeping his eye on the dog, and alert for the first sign that the dog may get up or break

FIGURE 26.—The command DOWN and STAY on leash.

away. (If there is evidence that the dog is not going to stay, the trainer goes back to the dog, motions him down, and repeats the command STAY, STAY.) He then walks back toward his dog, passes on his left side, walks behind him, turns around, and stands in position with the dog lying at his left side. During this movement, the trainer must repeatedly command: STAY, STAY, in a soothing voice. Then he gives his dog the command SIT, and when the dog obeys, praises and pats him.

(2) After this exercise has been successfully completed, by all trainers under the supervision of the instructor or assistant instructor, the group is ordered to march around the training square with the dogs at heel, before being lined up for a repetition of the exercise. Each time this exercise is repeated the instructor or assistant instructor should allow a slightly longer interval to elapse before sending the trainers back to their dogs.

b. Stay (sit).—Following the same procedure the trainer will train the dog to stay in a sitting position (fig. 27), and will alternate prac-

FIGURE 27.—The commands SIT and STAY on leash.

tice in this exercise with practice in the down position until the dog is able to stay in a sitting position.

c. Silent tie-up.—The trainer fastens the dog to a post with a light chain. He then walks away a short distance, remaining in sight of the dog. If the dog starts to whine or bark, the trainer reproves him with a sharp "No," comes up to him and quiets him. When the dog has remained quiet for about a minute, the trainer comes back, and unfastens and rewards him. The length of time that the dog remains tied is gradually increased. As the dog adjusts himself, the trainer goes out of sight. Later the leash is substituted for the chain and the trainer stations himself where he can watch closely, for if the dog once bites through a leash, he may become a confirmed "leather cutter."

Section IX

COME (ON LEASH)

118. Standard of performance.—When the dog has learned to obey the command Stay and is in "stay" position, the trainer calls the dog's name, immediately adding the command Come. The dog must promptly come and sit in front of, and facing, the trainer. At the command Heel the dog goes to the left side into the heel position.

119. Training procedure.—*a. Initial exercise in "come."*—The trainer commands: DOWN—STAY. He then backs away to the end of the leash. After the dog has obeyed the command and has settled in the down position, the trainer calls the dog's name, followed by the command Come, at the same time tugging slightly on the leash to suggest the meaning of the command. He repeats Come. As soon as the dog reaches the trainer, he gives the command: SIT, helping with his hand, if necessary, to make the dog sit directly facing him. The trainer next gives the command: HEEL and makes the dog come to heel position, then pets and praises him. This exercise is repeated until the dog is able to perform it without having the trainer tug on the leash.

b. Advanced exercises.—A 25-foot long leash is substituted for the standard leash. The trainer proceeds as before, gradually backing away from the dog until he has reached the end of the leash. He calls the dog, preceding the command Come with the dog's name, making sure that he holds the dog's attention. If the dog starts sniffing the ground or turns away to watch another dog, the trainer immediately jerks the leash and repeats the command Come in a stern voice; if necessary, he may pull the dog in all the way. After a few days, all of the dogs in the group will be able to carry out the command without difficulty. The instructor or assistant instructor then separates the trainers by 50 or 60 feet and instructs them to proceed, as follows: Permit the dog to wander around at the full length of the leash, sniffing and smelling as he pleases. At intervals of 1 or 2 minutes, call the dog by name and give the command: COME. If the dog does not come promptly, tug on the leash. If necessary, jerk at intervals and induce by voice. As soon as the dog is within touching distance, pat and praise him and then allow him to wander away again. It may be necessary, at first, to encourage the dog to go off by playing with him and pushing him. As soon as the dog starts to wander away, indicate approval by the word "OK." The dog will come to realize that this is a signal of release; that it means he is on his own and can romp and sniff all he wants. In calling the dog, the tone of voice should be authoritative, but never cross. When the dog does come, he should be praised, never punished—even if he has been pulled on the leash for the entire distance.

Section X

CRAWL (ON LEASH)

120. Standard of performance.—At the command Crawl, the dog crawls alongside of or towards his master (fig. 28). Ability in

Figure 28.—Executing the command Crawl (on leash).

this exercise is necessary to insure fulfillment of a mission under enemy fire or observation.

121. Training procedure.—The leash is held closely by the trainer, who commands: DOWN and lies down himself. Then, simultaneously, he commands: CRAWL, starts to crawl himself, and encourages the dog to do likewise. The leash may be used to encourage or correct the dog. When this exercise is readily performed by the dog, the trainer undertakes another phase of the work. Standing erect and facing the dog at a distance of a few feet, he commands: DOWN and then CRAWL to get the dog to crawl to him.

When the dog becomes proficient in this exercise, the distance between dog and trainer is gradually increased.

Section XI

JUMP (ON LEASH)

122. Standard of performance.—Almost any dog can and will jump or scale a $3\frac{1}{2}$ foot wall. The objective of this training ex-

FIGURE 29.—Performing the command JUMP (on leash).

ercise is to get him to jump on command, and ahead of the trainer, which is something entirely different. At the command UP the dog must immediately jump or scale the wall and then heel, or return to the trainer, according to the trainer's position. (A dog jumps a low hurdle or fence, which means that he clears it. He scales a higher hurdle or wall, which means that he jumps as high as he can, then scrambles or climbs the rest of the distance in order to get over) (fig. 29).

123. Training procedure.—*a. Initial exercises.*—(1) A dog may be afraid of a hurdle, even though he may have been taught to jump hedges or other obstancles. Therefore, it is well to use the hurdle, removing all the boards except the two bottom ones so that the hurdle is low enough to walk over.

(2) The instructor or assistant instructor, instructs members of the group to heel their dogs around a small square. He directs the trainer at the head of the column to lead over the hurdle. Each man steps over the hurdle and, as he does so, gives his dog the command: UP. If the dog hesitates or balks, the trainer stays on the far side and coaxes and helps his dog over, tugging on the leash. As the dog lands on the other side, the trainer steps away from the hurdle, giving the command: HEEL. This exercise should be repeated as often as necessary to make every dog in the group comply with the command. Those dogs who learn more readily than others are kept in line, since it is desirable for backward dogs to see others go over the hurdle successfully.

b. Advanced exercise.—(1) When the entire group has succeeded in clearing the hurdle at its initial height, the hurdle is raised, one board at a time, and the jumping exercise repeated. By the time the fourth board is inserted into the hurdle, every dog should be so proficient in the execution of the command UP that it will be no longer necessary for the trainer to step over the hurdle. Each man passes around the right side of the hurdle with his left hand extended, holding the end of the leash, which should be loose. As he does so, he gives the command: UP.

(2) Exercises in jumping from a sitting position are also undertaken, as follows: The hurdle has three boards in place. Each trainee in turn walks up to the hurdle with his dog heeling on the leash. Stopping about two feet in front of the hurdle, he makes his dog sit; then giving the command: UP the trainer walks around the right side of the hurdle holding the leash in his left hand, which is extended to permit the dog to jump freely. As the dog lands on the other side, the trainer praises his dog, gives the command HEEL, and walks to his place in line. After every dog has gone over the three-board hurdle, a fourth board may be added and the exercise repeated. The dog is made to sit at a distance from the hurdle equivalent to the height of the hurdle.

124. Caution.—Practice in jumping should not be overdone in any period. Although they enjoy it, it is very tiring to the dogs. Other exercises may be practiced as alternates to jumping if the dog shows signs of fatigue. Jumping, properly taught, is an exercise which raises the dog's spirits, and can be used as an alternative to depressing exercises such as "down". Jumping should include a variety of obstacles such as hedges, ditches, water, and walls.

Section XII

OFF LEASH EXERCISES

125. Standard of performance.—The standard of performance in heeling and sitting while off the leash is identical with that for on-leash activity, except in the exercises described on the following pages. The chief difference is that the physical means of controlling the dog has been removed.

126. Training procedure.—*a. Organization.*—When the instructor or assistant instructor observes that the dogs in his group are working dependably on a loose leash, he knows that the dogs are ready for training off the leash. To begin with, the instructor or assistant instructor selects the dogs which are doing the best work in his group. The dogs selected form a special class in off-leash training. Other classes are added until finally all dogs are working off the leash. Dogs that do not give a dependable performance are given extra training periods so as not to delay the training of the other dogs.

b. Methods.—Training the dog to work off leash follows the same procedure as that described for working on leash. A dog is worked off leash when he is obedient and dependable in all basic exercises. During all off-leash work, the trainer depends on the tone of his voice and his gestures to direct the dog; and on the affection which has been established between trainer and dog, and the attitude of obedience learned by the dog. Nevertheless the trainer must be alert in anticipating any hesitancy on the part of the dog to comply with his commands and must be able to prevent his dog from breaking or getting out of position. Once a dog has found that he can break control, it will take a long time to make that dog dependable off the leash. When a dog does disobey or break away he should be put on the leash immediately and carefully worked again under such control until the trainer decides that it is safe to take him off the leash again.

c. Subsequent practice.—Even after the dogs have advanced to specialized work, each training period should begin with a few minutes exercise heeling off the leash. The ability to perform off the

FIGURE 30.—Executing the command CRAWL (without the master) off leash.

leash correctly and dependably is fundamental to all advanced training (fig. 30).

SECTION XIII

STAY (OFF LEASH)

127. Standard of performance.—At the command DOWN, STAY or SIT, STAY, which may be given at any gait, the dog must promptly assume the position called for while the trainer continues ahead (fig. 31). The dog must remain in this position until the trainer either returns or gives him the command: COME.

128. Training procedure.—*a. Prerequisites.*—There are two prerequisites to this training.

(1) The dog has learned to heel off leash.

(2) The dog has learned to respond to the command COME from the end of the 25-foot leash.

b. General precautions.—Careful control of the dog by voice and general attitude is essential.

c. Initial exercises.—These follow the same procedure for the STAY exercise on leash (par. 117).

d. The standing halt.—Exercises in responding to the command STAY during movement. The last step is to make the dogs stay in position on command given during walking exercises. With the entire group under the direction of the instructor or assistant instructor, the trainers start walking their dogs at heel. When signaled to do so, each trainer commands: STAY, reaches down and places his left hand gently on the dog's head to halt him. He repeats the command STAY, making a backward motion with his left hand, turning slightly to watch the dog (fig. 32). When he has proceeded about 10 feet, he turns around and faces the dog. After an interval, he returns to his dog and gives the command: SIT. This exercise is repeated several times. Periods of heeling are interspersed to break the monotony. As the dogs become more and more dependable, the trainers may increase the distance they walk after giving the command STAY. When all the dogs in the group can be made to stay in their positions while the trainers continue walking, they are ready for the last stage of the stay training.

e. Exercise in staying in position with trainers out of sight. Under the supervision of the instructor or assistant instructor, each trainer commands his dog: STAY DOWN and continues walking about 25 yards to a spot where a hiding place has been arranged. The scaling hurdle may be used for this screen, particularly since the trainers may look through cracks in the boards and watch their dogs unobserved. If a dog starts to rise or break, his trainer should come out from behind the wall, advance slowly to the dog, speak to him sternly, and put him back in place. He then returns to hiding. The instructor or assistant instructor uses his judgment as to the length of time the trainers are to hide. At the start, he waits only a few seconds before directing the

FIGURE 31.—Executing the command SIT and STAY off-leash.

FIGURE 32.—Executing the command STAY. (Note gesture.)

men back to their dogs. As the exercise is repeated, the interval of hiding is increased until the dogs will stay in either the sit, stand, or down position indefinitely with the trainers out of sight.

SECTION XIV

DROP (OFF LEASH)

129. Standard of performance.—With the dog in a stay position at a distance of 50 feet or more, the trainer calls the dog's name, followed by the command COME. Before the dog reaches him, he gives the command: DOWN or makes a sweeping downward movement with his arm or does both. The dog drops promptly, regardless of his gait, and stays in that position until a new command is given (fig. 33).

130. Training procedure.—*a. Prerequisite.*—Group work in this exercise is started only when all the dogs in the group perform satisfactorily in the "stay" and "come" exercises off the leash.

b. Organization.—The trainers and dogs are in line facing the instructor or assistant instructor, with a distance of about 10 or 15 feet between trainers. At the direction of the instructor each trainer faces his dog and proceeds as outlined below.

c. Methods.—(1) The trainer gives the dog the command: DOWN and STAY, turns and walks away to a distance of about 25 feet, and turns again so as to face the dog. He calls the dog's name and follows it with the command COME. When the dog is within 10 feet of him, he gives the command: DOWN in a firm voice, accompanied by a downward motion of the arm. As he does this, he takes a step or two forward toward the dog. At first, he may have to repeat the command and the arm motion several times. It may even be necessary to approach the dog and make him assume the down position.

(2) After the dog has dropped into the down position, the trainer calls the dog by name and gives the command: COME. When the dog has obeyed, he gives the command: HEEL and walks the dog back to the starting point to repeat the exercise. After the dog has learned to drop instantly on command, the trainer varies the distance between himself and the dog before giving the command. As the dog becomes more and more proficient and dependable in this exercise, the trainer uses the arm signal and the verbal command independently so that the dog learns to respond to either.

(3) This exercise is depressing to the dog's spirits and must not be repeated too often in succession.

FIGURE 33.—DROP. Executing the DOWN command by gesture—off-leash.

SECTION XV

JUMP (OFF LEASH)

131. Training procedure.—When all the dogs in the group have demonstrated ability to take the full sized hurdle while on the leash, and to carry out commands off the leash, they are ready for training in jumping over or scaling hurdles and other obstacles while off the leash (fig. 34.) The hurdle is gradually heightened. When the dog is able to get over the hurdle at its full height and to return to position as commanded, practice may be undertaken in jumping over or scaling other obstacles. Practice in these exercises should not be overdone in any one period. Advanced exercises in jumping may include variations in all basic exercises (fig. 35).

FIGURE 34.—Performing the command JUMP off-leash.

SECTION XVI

ACCUSTOMING TO MUZZLES AND GAS MASKS

132. General.—Both muzzles and gas masks are put on the dogs after they have learned to heel on leash. The equipment should fit the dog snugly but comfortably.

133. Muzzles.—When the muzzle is adjusted, the trainer starts heeling the dog at a fast walk on leash. If the dog tries to reach up with his front feet and pull off the muzzle, the trainer jerks upward

FIGURE 35.—Variation in scaling off-leash, demonstrating discipline.

with the leash and keeps on walking, holding the leash taut. As soon as the dog ceases his attempts to remove the muzzle, the leash is slackened and the dog is praised.

134. Gas masks.—The dog is accustomed to the muzzle before a gas mask is put on him. The first time the mask is used, he must not be moved so quickly as to cause him to pant. Accustoming the dog to this new equipment depends on patience and practice.

SECTION XVII

CAR BREAKING

135. Standard of performance.—A dog can be considered car broken when he enters a motor vehicle on command, stays quietly within, and makes no attempt to get out unless ordered to do so (fig. 36). In exceptional cases, dogs never become accustomed to riding in motor vehicles. Such dogs should be considered unfit for training for military purposes.

136. Training procedure.—*a. Initial exercise.*—(1) To get the dog to enter a vehicle and stay there, the commands UP and STAY are given. The vehicle is then started. Upon noticing the first symptoms of car sickness (a profuse flow of saliva and drooling from lips and tongue, followed by gagging), the instructor decreases the speed of the vehicle and observes whether symptoms continue. If the dog shows improvement, speed can be gradually resumed, and this procedure repeated upon lessening of symptoms until the dog becomes accustomed to the motion of the vehicle.

(2) If drooling continues when speed is diminished, the car is stopped and allowed to remain stationary until the dog's behavior is normal. The tendency to be sick can sometimes be checked by the instructor's speaking to the dog. The vehicle can then be put into motion again, and this procedure repeated as necessary.

(3) In cases where the dog has exhibited symptoms of car sickness, it will be found advisable to break him to riding when his stomach is empty.

b. Advanced exercises.—(1) In subsequent exercises the length of the rides must be progressively increased—5 minutes the first day, 10

FIGURE 36.—These dogs are car-broken. They present no problem in transporting.

minutes the second, and so on until the dog shows no evidence of car sickness, no matter how far he travels. The speed of the vehicle should also be varied.

(2) After it has been demonstrated that the dog can ride without becoming car sick, the same procedure should be followed with the dog loaded in the car in a crate.

Section XVIII

ACCUSTOMING DOG TO GUNFIRE

137. Standard of performance.—Dogs may be considered inured to gunfire when after training it is possible to discharge weapons of small caliber directly over the dog without causing him to flinch. Some few dogs can never be trained to be steady under fire. Such dogs should be considered unfit for training for military purposes.

138. Training procedure.—Dogs can be accustomed to gunfire by discharging weapons of small caliber from a distance, gradually moving gunfire nearer to the dog and firing progressively heavier caliber shot. It is best to start this training when dog is engaged in some activity which absorbs his attention, such as eating, or during training periods. Thus the dog subconsciously becomes accustomed to distant gunfire and firing can be resumed closer and closer without unduly disturbing him. It has been found that the most successful procedure in inuring a dog to gunfire is to order shooting done as casually and intermittently as possible. When dog shows alarm, no attempt should be made to force him to sit quietly while firing is repeated. It is better to discontinue firing for the time being and resume when dog has forgotten his alarm.

Section XIX

BASIC TRAINING RECORDS

139. Trainers.—*a. Importance of records.*—The maintenance of records pertaining to the qualifications and efficiency of each trainer is of paramount importance. These records should contain sufficient information to make possible the proper utilization of training personnel. Some trainers may prove satisfactory during initial stages of training and subsequently, for some reason or another, may fail in their job and have to be disqualified. Other trainers may prove satisfactory only for purposes of basic training. Still others may be assigned responsibilities for advanced training of dogs, and the nature of these assignments must be determined by available information concerning their qualifications and efficiency.

b. Recommended forms.—To facilitate such classification and to insure standardization in procedures, the record form (p. 90) may be used throughout the period of basic training. The information test referred to in the standard personnel record form is given on pages 89–90. It is administered at the end of the first two weeks of training, which consist of lectures and demonstrations as well as introduction to actual work with dogs. The practical tests require simply that each trainer shall go through the various on-leash and off-leash exercises with each of his dogs. His rating depends on the degree to which trainer and dog meet the standard of performance that has been set for each exercise.

c. Superior trainers.—Trainers rated as superior may, after successful completion of the advanced training program, be considered for positions as assistant instructors of future trainers.

140. Dogs.—Dogs cannot be rated haphazardly any more than trainers. Their performance must be carefully studied and numerous factors taken into consideration before finally being passed, or failed for on-leash and off-leash work. The record form on page 90 is particularly helpful as a guide to classification of basically satisfactory dogs for various types of specialized training.

(SAMPLE)

PERSONNEL RECORD (BASIC TRAINING)

Name of Student Trainer J. Brown Name of Instructor Everett Fisher
Selected for training after interviews by Lt. Virgil Kaufman
on (dates) 1–20–43; 1–25–43

INFORMATION TEST

Date 2–2–43 Rating (SS for superior; S for satisfactory; U for unsatisfactory) S

PRACTICAL TESTS

On-leash exercises	*Off-leash exercises*
Date of test 2–13–43	Date of Test 2–19–43
Rating (SS, S, or U) S	Rating (SS, S, or U) SS
Examiner E. Fisher	Examiner E. Fisher

REMARKS

As occasion arises, note any significant observations on the following qualifications # of the trainer (place initials and date after each observation):

Attitude toward dogs: Very fond of the dogs he has been training (E. F. 2-11-43)

Intelligence: About average (E. F. 2-5-43)

Patience and perseverance: Generally satisfactory but became irritated once (E. F. 2-6-43)

Coordination: Good (E. F. 2-8-43)

Physical endurance: Good (E. F. 2-7-43)

Resourcefulness: Fair. Follows rules literally regardless of situation (E. F. 2-10-43)

Dependability: Good. Realizes that dog's welfare is in his hands (E. F. 2-5-43)

Other observations: Inclined to be overindulgent but has learned importance of firmness (E. F. 2-7-43)

GENERAL RATING

On the basis of the above information and his personal observations, the chief instructor or some other authorized individual should rate the trainer as (check in proper place):

Superior—— Satisfactory X Unsatisfactory——

Name of rater Lt. Kaufman Date 2-20-43

(SAMPLE)

DOG TRAINING RECORD (BASIC)

Name of Dog DUKE Trainer(s) JOHN BROWN: WILL GREEN

Breed Airedale Terrier Sex Male Height 23 in.

Color and Markings Black and Tan Brand D001 Weight 55 lbs.

Characteristics of dog observed during basic training period	Degree shown—	
	Appraisal at end of on-leash training	Appraisal at end of basic training
Intelligence	H	H
Willingness	M	H
Energy	H	H
Aggressiveness	L	M
Sensitivity	M	M

John Brown John Brown

Signature of rater:

NOTE: Write "H" for high degree; "M" for a moderate degree; "L" for a low degree of the trait in question.

REMARKS: No difficulties were experienced in the training of this dog.

(Use reverse side if necessary)

GLOSSARY:

Intelligence: A dog's rating in intelligence is based upon the readiness with which he learns and the extent to which he retains and uses what he has learned.

Willingness: A dog is ranked high in willingness if he consistently responds to his master's commands with an effort to carry them out, regardless of any prospect of reward or punishment.

Energy: This term refers to the degree of spontaneous activity by the dog; in other words, to the speed and extent of his movements in general, not in response to any command.

Aggressiveness: This term is to be considered literally. It refers to the degree to which the dog asserts himself or is inclined to attack.

Sensitivity: A dog is rated over-sensitive if he is easily distracted and timid. He is rated as under-sensitive if he has limited perception and is inclined to be stolid and indifferent. Dogs of moderate sensitivity are usually desirable for training purposes.

EXAMINATION ON WAR DOGS

(To be given when the student trainer has been taught chapters 1 and 2)

SECTION A

Place the letter "T" before each of the following statements that is true and the letter "F" before each false statement.

-- 1. Dogs were not used for military purposes until World War I.
-- 2. Dogs can detect slight movements of objects and persons.
-- 3. Dogs have a much keener sense of smell than humans.
-- 4. The German and Japanese armies are making extensive use of war dogs in World War II.
-- 5. Canine vision is superior to human vision.
-- 6. There is no one breed to be preferred over every other breed for military purposes.
-- 7. When in the field, there is no harm in permitting the dog to share the quarters of his handler.
-- 8. "Dew claws" are of no concern in the care of a dog.
-- 9. Crosses of various breeds are not desirable for military training.
— 10. A keen sense of smell is essential for all military functions and facilities training in such functions.
-- 11. Dogs may be fed meat that is raw, boiled, or roasted.
-- 12. Kennels should not be placed face to face.
-- 13. A mediocre trainer may spoil a dog that is well equipped for military training.
-- 14. Dogs are more sensitive to sound than humans.
-- 15. A dog should be permitted to drink as much as he wants to.
-- 16. Bones are essential to the diet of a grown dog.
-- 17. A normal healthy dog should be bathed as infrequently as possible.
-- 18. A dog must never be fed his main meal just before he goes to work.
-- 19. A dog that is undersensitive in hearing will also be undersensitive to touch.
-- 20. Regardless of the nature of the injury, an injured animal should be moved without delay to a place where adequate care can be given.

SECTION B

Fill in the missing word or words in each of the following statements.

21. Responsibility for the training of war dogs is specifically delegated to________
_______________________________.

22. The two types of training provided for war dogs are known as:

    (a) ______________________________.
    (b) ______________________________.

23. The type of dog most likely to be successful in military training is___________ sensitive to sound and __________ sensitive to touch.
24. Ordinarily the number of meals a day sufficient for a mature dog is__________.
25. In brushing a dog's coat, the correct procedure is to brush it first _________ the grain.
26. List below three places where individual kennels are used:

    (a)______________________________.
    (b)______________________________.
    (c)______________________________.

27. A dog that is rated under __________ cannot be taught to attack.
28. The trait most clearly related to a dog's success in military training and service is ________________.
29. A dog's willingness can be __________ or _______________ by the man who handles the dog.
30. In attempting to eliminate fleas from the dog, it is necessary to disinfect the dog's bedding and kennel because_________________________________
____.
31. When a dog is moved from one kennel to another the old bedding should be ________________.
32. In case of injury to a dog, or illness, the assistance of the _________________ is to be sought.
33. Besides meat, other sources of protein are:

    (a)_______________ (b)_______________ (c)_______________.

34. Meat or other sources of protein should constitute about _________________ of the total ration for the dog.
35. Cereal grains most commonly used in some form for dog feeding are: corn, rice, and (a)_______________(b)_______________(c)_______________.
36. One of the tactical services for which dogs are trained is ___________________.
37. Simple abrasions of the skin should be washed with ___________________ ______________________if no suitable antiseptic is available.
38. The _______________ should always be included when bandaging the leg of a dog.
39. Because of the possibiity of ____________, anyone bitten by a war dog is required to report his name and the ____________ of the dog, no matter how small the wound.
40. In case of shock resulting from a fall, collision with a vehicle, or contact with electricity, the dog should be kept _______________ and quiet until veterinary aid is obtained.

SECTION C

Answer each of the following questions in the manner specified.

41. Number in proper sequence the following procedures that are carried out when a dog arrives at a reception and training center.

-- The dog is equipped with a leather collar carrying a metal tag.
-- The dog's number is tattooed in his left ear.
-- Data are obtained for the record card, Form 120.
-- The dog is exercised on a leash.
-- The dog is left alone in the kennel, with food and water, for a period of time.

42. Fill in a symptom or characteristic that is indicative of an undesirable trait or condition:

(a) Forefeet____________________.
(b) Pads____________________.
(c) Body____________________.
(d) Breath____________________.
(e) Mucous membrane____________________.

43. Below are a list of breeds and a list of special traits. Match both lists by writing the number of each breed on the blank line preceding the appropriate trait. In some cases, more than one number may appear on the same line.

1. Labrador Retriever
2. Chesapeake Bay Dog
3. Alaskan Malamute
4. Boxer
5. Dalmatian
6. Eskimo Dog
7. German Shepherd
8. Standard Poodle
9. Samoyede

-- "Snoeshoe" foot.
-- Occasional deafness in the breed.
-- White or cream-colored, weather-resisting coat.
-- Exceptional swimming ability.
-- Rapidly growing coat.
-- Brilliant markings that usually have to be camouflaged.
-- Wolflike appearance.
-- Undershot jaw.

NOTE—IN SCORING A STUDENT'S PAPER, GIVE ONE POINT FOR EACH CORRECT ANSWER IN SECTION A, TWO POINTS FOR EACH CORRECTLY FILLED BLANK IN SECTION B, FIVE POINTS (OR A DUE FRACTION THEREOF) FOR EACH OF THE FIRST TWO QUESTIONS IN SECTION C, AND TEN POINTS (OR A DUE FRACTION THEREOF) FOR THE LAST QUESTION IN SECTION C.

Total number of possible points in Section A	20
Total number of possible points in Section B	60
Total number of possible points in section C	20
Maximum score	100

A SCORE OF SEVENTY SHOULD BE CONSIDERED PASSING.

CHAPTER 4

SPECIALIZED TRAINING

SECTION I

SCREENING FOR SPECIALIZED TRAINING

141. General.—The term "screening" refers to the process whereby dogs are initially classified for the type of military service for which they seem best qualified. This classification is based on observation of the physical and psychological traits manifested by dogs under preliminary examination immediately after arrival at war dog reception and training centers. Future observations made during actual training may substantiate the results of the screening process or result in reclassification or rejection of dogs. It is imperative that such observations be made throughout the training program. The potentialities of some dogs and inherent weaknesses of others may come to light as suddenly as in the case of human beings. In general, however, the screening process will prove to be fairly reliable if the dogs under examination are classified with regard for the traits specified in this manual as essential for each type of war dog (pars. 146, 154, 161, 169, 177).

142. Recommended procedure.—The procedure recomemnded for initial screening of war dogs at reception and training centers is illustrated in figure 37. Newly received dogs are first examined by the veterinary service, which rejects those that are unfit for physical reasons. The dogs are then observed by a classification board consisting of the Commanding Officer and the veterinarian or their appointed representatives. On the basis of the physical and psychological traits manifested and required, the board may make further rejections and then assigns the remainder to the attack, messenger, scout, and casualty dog branches. Sentry dogs, which are the least specialized in function and require the least expenditure of time and effort, are drawn from dogs found unsuitable for training in the other branches, as indicated on the chart. Those dogs not retained by the sentry branch may be put in a pool for reclassification or final rejection

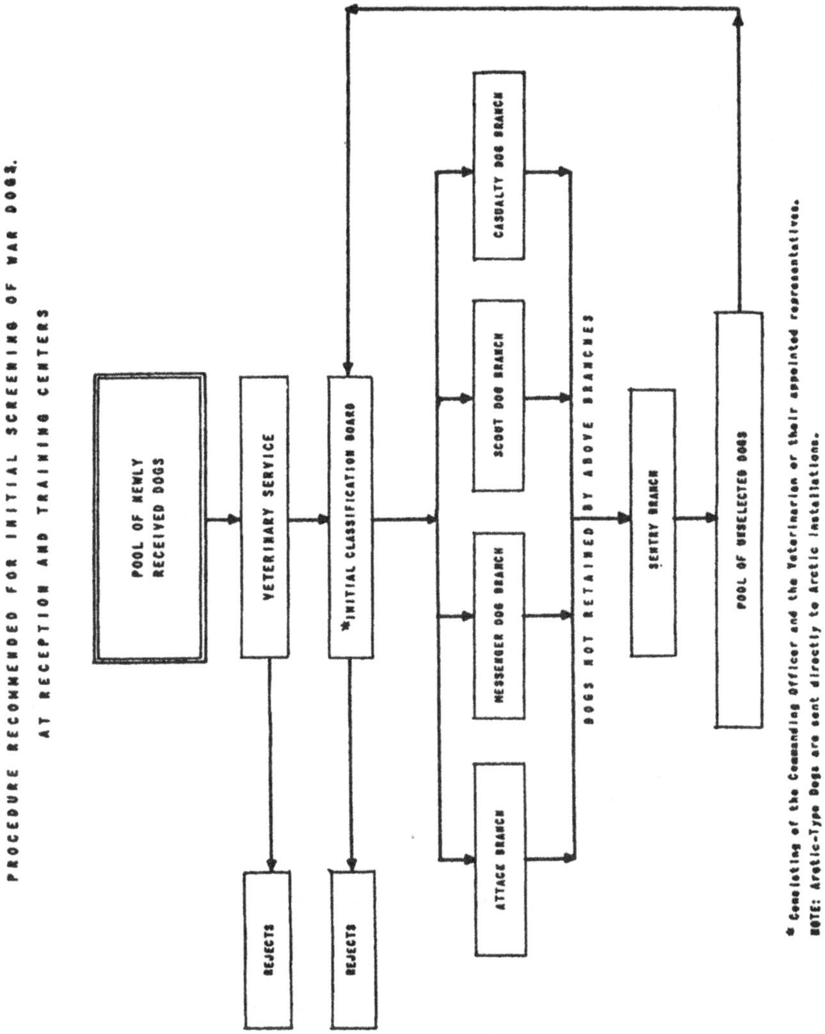

FIGURE 37.—Initial screening of war dogs.

by the classification board. This screening process may extend over a period of several days, during which time the instructors in each branch make appropriate observations concerning the potentiality of all dogs routed to them. The entire procedure, however, may be modified in accordance with current requirements for specific types of war dogs and available supply of dogs for all training purposes.

SECTION II

PRINCIPLES OF SPECIALIZED TRAINING

143. General.—Basic training of dogs, as an end in itself, has no place in the military program. It serves simply, but necessarily, to develop in dogs the behavior that is essential to efficient and effective training for specific military functions. The dog that has successfully completed basic training is disciplined, prepared to absorb instructions of a specialized nature. Furthermore, his behavior during basic training, if properly observed, is an indication of the type of specialized training for which he may be best suited. Basic training of enlisted men serves similar purposes. In specialized training, general finish and discipline are subordinated to the successful accomplishment of specific missions.

144. Basic principles.—The effectiveness of specialized training will depend on the regard shown for certain fundamental principles:

a. The dog must not be trained for more than one job. He can learn to be a good sentry, scout, messenger, attack dog, or casualty dog. Training for more than one job will, because of the natural limitations of the dog, diminish the contribution he can make. The job for which he is trained will depend on his basic traits and training. A screening process (pars. 141 and 142) is utilized to select the most suitable field of training for a given dog.

b. The general attitude of the student trainer is all-important. He must fully realize the significance of the work that he is doing. Dogs have been embraced for military service to conserve manpower, to safeguard life, and to further the work of the military service by utilization of powers that humans do not possess to the same degree as dogs.

c. The importance of the man-dog relationship cannot be overestimated. The dog and his trainer must work as a team. Consequently, no trainer must be charged to train an animal that he deems unsuitable. For the same reason, if a dog appears unwilling to serve a certain trainer it may be necessary to make some other arrangement. On the other hand, once a team has been established during a training period, the relationship must not be interfered with by anyone other than the trainer petting, feeding, or otherwise handling the dog.

d. Association of ideas greatly facilitates learning. Where special equipment is used, the dog must learn to associate this equipment with his work. Thus, the special message-carrying collar should be put on the messenger dog only when he is ready to be worked; it should be removed as soon as the training exercise is completed.

e. The dog should be motivated not only by praise and petting of his trainer, as in basic training, but also by the goal of accomplishment of a mission. The dog can and should be trained to complete a task as an end in itself, not simply for the sake of reward by his master. In all his training, therefore, he must be permitted to finish every exercise successfully, no matter how many errors he makes during the course of it. THE DOG MUST ALWAYS "WIN."

f. Specialized training should be so directed that the soldier comes to rely on the dog for the accomplishment of things that the soldier himself cannot do as well or as quickly. Eventually the soldier depends upon the dog, whereas in basic training the dog is always guided by the trainer.

g. In order to develop the dog's responsibility for given tasks and to assure the fulfillment of his mission in the theatre of operations, it is essential that specialized training be carried out over varying terrain and in the face of gunfire and other distractions.

h. The ability of the dog is fully developed by training during daytime. The effects of such training will carry over into actual service at night, when training is difficult to conduct because of darkness, which interferes with careful observation and corrective measures. The effects of training during daytime will be enhanced by the increased sharpness of some of the dog's powers during nighttime.

i. Reviews of previous training exercises are essential to maintain and raise the level of performance. Trainers must use their best judgment in scheduling, and determining the extent of reviews.

j. Successful training of war dogs depends on the care taken for their welfare. Unless the dogs are kept in good health, properly groomed, fed, and kenneled, the effectiveness of the training program will be diminished.

SECTION III

DOGS FOR INTERIOR GUARD—THE SENTRY DOG

145. Use of sentry dogs.—The sentry dog, as the name implies, is used principally on interior guard duty as a watch dog (fig. 38).

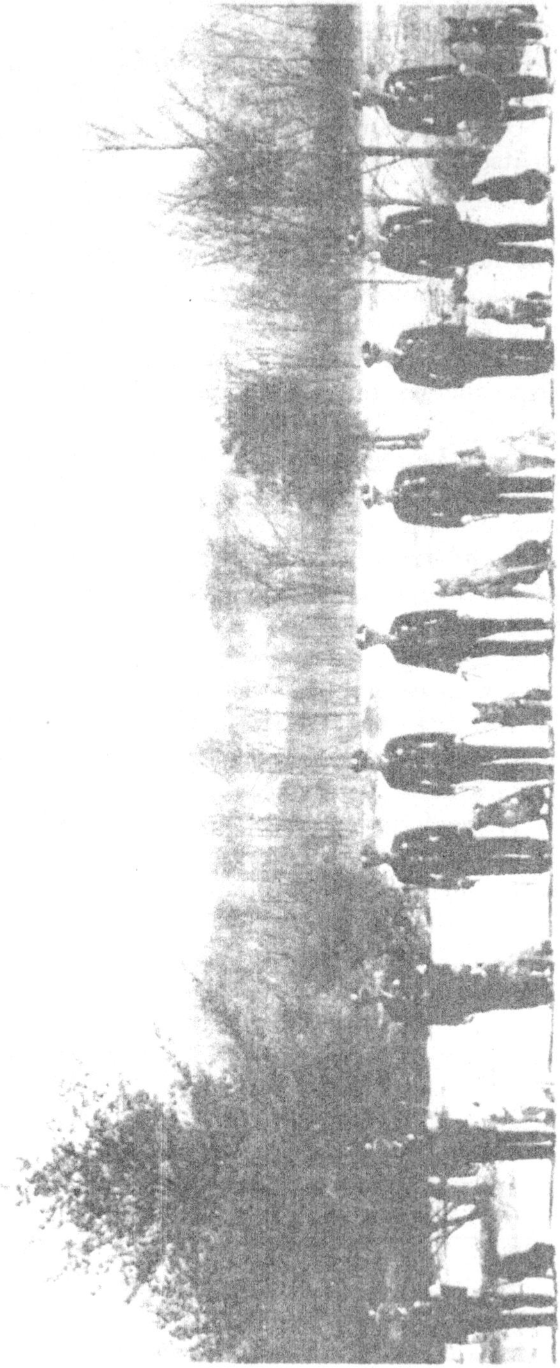

FIGURE 38.—Sentry dog protective unit ready to go on duty.

This class of dog is trained to give warning to his master by growling or barking or silent alert. He is habitually worked on leash. The sentry, keeping the dog on leash, patrols his post in daylight or darkness, and can depend upon the dog to notify him of the approach or presence of strangers within the confined area being protected. After

FIGURE 39.—The sentry dog alerts his sentry.

the dog alerts (fig. 39), the sentry thus alerted must be prepared to cope with the situation as circumstances then dictate, that is, challenge; investigate, keeping carefully under cover; or make an arrest. The dog, being kept on leash and close to the sentry, will also assist as a psychological factor in such circumstances. He will even attack if an intruder should threaten the sentry, although he is not trained in the more advanced types of attack work. The sentry dog can be used to advantage in such critical areas as—

Plane dispersal areas.	Ammunition dumps.
Beach patrol areas.	Motor parks.
Gun emplacement areas.	Ration dumps.
Dynamite storage areas.	Waterworks.

Where areas to be guarded are free from traffic, entirely confined, and not excessively large, a specifically trained dog may be turned loose and permitted to roam at liberty. Only the dog's master should be permitted in or near the area under such guard and he must accompany any patrol sent to threatened points upon warning of trespass. Such dogs working at liberty are usually sharp and will attack trespassers at once. Sentry dogs may be used by the average individual after a reasonable period of instruction. It is important, however, that sentries be definitely assigned to each dog and changes thereafter be kept to a minimum. Best results are being obtained where two sentries are assigned to a dog. Being on duty from sundown to sunup, each sentry performs approximately an 8-hour shift. This unit works two nights and lays off one, thereby allowing both men and dog sufficient time for rest. Three such units are thus required for each two posts mounted.

146. Selection of dogs for training in sentry work.—Dogs for sentry duty, as indicated in figure 37, will be selected from those dogs which have been considered unsatisfactory for any of the four other classifications (attack, casualty, messenger, scout). However, to be suitable for this specialized training, dogs should possess the following traits (see pars. 14 to 18):

Intelligence:	Moderate to high.
Willingness:	Moderate to high.
Energy:	Moderate to high.
Aggressiveness:	Moderate to high.
Sensitivity:	Low to moderate.

Under certain conditions, dogs may be assigned for sentry duty without undergoing such screening. These conditions are to be determined by the commanding officer in the light of the demand for sentry dogs and that for other specially trained dogs.

147. Personnel.—*a. Function of trainer.*—In addition to teaching the dog to obey his commands, the trainer must instill in his dog the idea that every human, except himself, is his natural enemy. This he does by encouraging the dog to alert at the presence of any stranger and to attempt to attack him. The sentry dog does not actually attack; he is always worked on a leash. It must be emphasized that *the trainer never permits anybody to pet or make friends with any dog he is training*. He must remember that he is the only friend the dog must recognize; he is *the only master*. For the same reason, *he must never pet any dog except his own.*

b. Functions of the assistant.—An assistant, known as the aggravator, simulates an enemy. It is the function of the aggravator to sharpen the dog's aggressiveness and to build up his self-confidence by retreating as soon as the dog makes an aggressive move toward him. He approaches the dog in a furtive and threatening manner; at the slightest move by the dog toward him, he retreats. Latitude is permitted in the action taken by the aggravator, due to the fact that each dog requires different treatment. It is essential, however, that in every encounter between the dog and the aggravator, the DOG MUST ALWAYS BE THE WINNER. Different individuals are used as aggravators in order to make the dog generally antagonistic toward strangers. The aggravator's function is as important in sentry dog training as the function of the master and, therefore, his work should be carefully supervised by the instructor. The master of one dog can act as aggravator in helping the master of another dog.

148. Conditions for training.—*a. Location.*—Training starts in the regular training area. As the dog progresses, training continues in different locations, chosen to resemble various types of sentry posts.

b. Hours.—Initial training is done during daytime. When the dog alerts to every unfamiliar presence in daylight, final training is undertaken at night.

149. Equipment.—The following equipment is required:

a. Small flexible stick, rolled up sacking, or other harmless weapon.

b. Chain choke collar.

c. Six-foot leather leash.

d. Flat leather collar (fig. 40).

e. Stake.

f. Spring.

g. Chain leash or chainette.

h. 25-foot leash.

150. Basic training required.—*a.* Specialized training for sentry work is based on training in the following exercises (see ch. 3):

(1) Heel (close, loose).

(2) Down.

(3) Stay.

(4) Sit.
(5) Come.
(6) Jump.
(7) Standing halt.
(8) Cover.
(9) Crawl.

b. In addition, dogs to be trained for sentry work must be accustomed to gunfire and car-broken.

151. Training procedure.—*a. General.*—Training for the sentry dog is divided into two stages:

(1) *Testing, arousing, and controlling the dog's natural aggressiveness.*—Some dogs are naturally aggressive and do not need much

FIGURE 40.—Special flat leather collar for sentry, attack and silent scout dog.

teasing to become excited and pugnacious. In order to avoid accidents, sharpening or aggravating may be initiated by chaining the dog securely to a stake. A spring is used in fastening one end of the chain to the stake; the other end is attached to the broad leather collar. In this way the dog cannot hurt or choke himself when lunging at his tormentor. Where this spring is not available, it has been found advisable to loosen the earth around the top of the stake in such a manner that the stake will give in response to the dog's lunging and not check him too charply.

(2) *Teaching the dog to detect a stranger's presence or approach under any conditions and to warn his master.*—The dog must learn to detect a foreign object or presence either below, at, or above ground level. For this reason, the final 2 weeks of training must include work with the dog patrolling a given area in which the aggravator is concealed in such places as the branches of a tree, behind a fence, or in a ditch. It is desirable that the dog work on the 25-foot leash, as well as the short leash, so that he may be able to enter a suspected building or other place of concealment ahead of the master.

b. Specific.—Training proceeds as follows:

(1) If a stake is used, the trainer chains the dog to the stake and, in addition, puts on the chain-choke collar with the 6-foot leather leash attached. The trainer heels the dog to the end of the chain, orders him to sit, and steps away from the dog, keeping hold of the leash.

(2) The master puts the dog on the alert by the command WATCH HIM uttered in a low voice, almost a whisper. The command WATCH HIM serves only to put the dog on the alert during early training. It is a signal to the dog that he is on duty and must be prepared to detect any intrusion. The command should be eliminated as soon as the dog has learned that the putting on of collar and lead signifies "On duty." WATCH HIM is never used in actual service. The dog notifies the sentry of danger; the sentry does not notify the dog.

(3) The aggravator appears, equipped with a small, flexible stick, or other harmless weapon. He approaches the dog in a threatening manner, not facing him directly, but approaching from an angle. His manner indicates apprehension. He looks at the dog out of the corner of his eyes; he does not stare at him. He strikes at the dog without actually hitting him, and jumps away. As he strikes, the dog's master encourages the dog by saying, "Get him, get him," in a sharp voice.

(4) Dogs will respond to this procedure according to their natural aggressiveness. They may be conveniently classified as follows, and training continued accordingly:

(*a*) *The moderately aggressive dog.*—This dog is the easiest to train for sentry work. He barks or growls as soon as the aggravator appears, lunging at his chain and attempting to attack the aggravator. As soon as he responds in this manner, the trainer praises him lavishly, and, if the dog is not too excited, he steps close to him and pats him. When the trainer is convinced that the dog is aroused only against the aggravator and that he has no tendency to attack indiscriminately, he may be considered ready for work off the stake. The trainer then unchains the dog and the exercise for that day is concluded with the aggravator running away and the master praising the dog.

(*b*) *The overaggressive dog.*—This dog is apt to lose his head, attempt to bite anyone within reach, and keeps on barking even after the aggravator disappears. In this case, the master reproves the dog. He shouts, "No, no," jerking simultaneously on the leash until the dog quiets down. He must be careful not to step within reach of the dog until he stops barking and growling. When he is sure that the dog has quieted down sufficiently, he approaches him, speaking soothingly, and praises and pets him. This procedure is repeated, except that the aggravator appears and disappears immediately. As the dog shows signs of directing his aggressiveness toward the aggravator, the master lavishes praise on the dog, encouraging him to understand that his hostility must be directed towards the aggravator. After two or three trials, training stops for the day. It is resumed the following day with the dog chained to the stake as before. He is kept chained to the stake until he has demonstrated that he will not bite his master, no matter how excited he becomes.

(*c*) *The underaggressive dog.*—This dog reacts negatively to the presence of the aggravator. He may stand and wag his tail, throw himself on the ground, or try to run away from the aggravator. In this case, the aggravator approaches the dog from the flank and hits or seizes him from behind; the trainer encourages the dog, making threatening gestures toward the aggravator, who imitates a dog's growl. When the dog, in trying to protect himself, snaps or growls at the aggravator, the aggravator at once ceases to tease the dog and quickly steps away. It is very important in training this type of dog that the master exaggerate his praise and encouragement whenever the dog shows the slightest sign of aggressiveness, and that the aggravator exaggerate his simulated fear of the dog. This procedure must be repeated until the dog's confidence is built up and he attempts to attack the aggravator as soon as he approaches. The master then unchains him from the stake and training for the day is concluded with praise by the master and disappearance of the aggravator.

(5) Six naturally aggressive dogs are brought out at the same time. These are dogs which have demonstrated that they do not need to be chained to the stake. Between every two of these dogs is placed a dog who has reacted negatively to the first phases of training. They are lined up far enough apart so that they cannot get into a fight among themselves. Each dog is on leash at the left side of his master. At the command WATCH HIM, the aggravator appears and walks towards the dogs. Some of the dogs will bark immediately; these should be praised lavishly by their masters. The aggravator concentrates his attention on the dogs that do not respond readily. He approaches them with his stick, threatens them, and jumps away. Inspired by the bolder dogs beside them, even the slow ones will eventually start barking. If properly encouraged by their masters, they will understand that there is nothing to fear from the aggravator, and that he will disappear as soon as they bark, growl, or make a move toward him. When all the dogs in the group alert as soon as the aggravator appears, he must vary his direction of approach and increase the distance at which he first appears. The dogs that detect him earliest are praised

lavishly. It will be found that the slower dogs learn from their aggressive companions, as well as from their masters.

(6) When all the dogs in the class alert at the approach of the aggravator, a new man takes his place. The dogs learn in this way that *any* man approaching is an enemy. It is desirable to have numerous persons play the role of the aggravator.

(7) The master now plays the part of the sentry, walking post with his dog heeling on loose leash. (This simulated post must be changed every day so that the dog does not get accustomed to one definite route.) When they have advanced a short distance, the aggravator approaches furtively from some place of concealment, such as trees or tall grass. If the dog has learned his first lessons, he will detect the approach of the aggravator and will alert without help from the master; but if he fails, the master helps him by saying, "Watch him." As soon as the dog gives warning, the aggravator runs out of sight, the master simultaneously praises and encourages his dog. In the event that the dog fails to respond to the aggravator's approach, the aggravator conceals himself along the path of the master, steps out quickly from his hiding place, hits at the dog with the stick, and jumps away. This will arouse the dog. Furthermore, he will learn that unless he gives alarm immediately upon detecting the presence of a stranger, his lot will be pain and punishment.

(8) When a dog detects any strange presence at a considerable distance without any help during the daytime, he is generally ready to be worked at night. It will usually be found that a dog will work better at night, as scenting conditions are more favorable and his keen hearing is enhanced by the absence of distracting noises.

152. Additional training notes.—*a. Importance of loose heeling.*—In walking a post, loose heeling is essential as long as the dog does not pull or tug on the leash. If close heeling is insisted upon, the dog is apt to concentrate on perfection in heeling; this means that his attention will be riveted on the master and not concentrated on his environment. He is apt to forget his main duty—to be on the alert at all times and ready to give alarm at the slightest provocation.

b. Importance of distrust of strangers.—The sentry dog is taught not to make friends with strangers. To insure such distrust, the following method may be employed: The master takes his dog on leash at the heel position. A stranger approaches uttering ingratiating words and coaxes the dog to come to him. As soon as the dog makes an attempt to respond, the stranger slaps him smartly on the nose and jumps away. The master then encourages the dog to attack the stranger. This is repeated until the dog growls and barks on the approach of the stranger, no matter how friendly his attitude or how much he attempts to appease the dog. The stranger then tries to coax the dog away with some choice tidbit, meat, or anything else of which the dog may be particularly fond. If the dog attempts to take the food from the stranger's hand, the stranger again slaps him on the nose and runs away without giving him the food. Thus the dog learns that HIS MASTER IS THE ONLY ONE TO BE TRUSTED.

SECTION IV

DOGS FOR INTERIOR GUARD

THE ATTACK DOG

153. Use.—*a.* The attack dog is trained and used to supplement the sentry dog. He attacks off leash on command, or on provocation, and ceases his attack on command or when resistance ends. He may be used in connection with, or in place of the sentry dog in situations where it is necessary to—

(1) Apprehend a malefactor at a distance from the sentry.

(2) Replace a sidearm in congested areas where it would be dangerous to attempt to shoot at a fleeing malefactor.

(3) Guard prisoners.

(4) Transport prisoners.

b. Where the attack dog is to be used in place of or to supplement a sentry dog, the dog's master must handle him. It may also be practical to use attack dogs with scout dogs, for example, in dense jungle areas or under certain circumstances at night. The tactical situation is the chief determinant of manner of use.

154. Selection of dogs for training in attack work.—*a.* To be suitable for this specialized training, dogs should possess the following traits:

(1) *Intelligence.*—High.
(2) *Willingness.*—High.
(3) *Enegry.*—High.
(4) *Aggressiveness.*—High.
(5) *Sensitivity.*—Low to moderate.

b. The attack dog must also possess courage, speed, and strength. He must be medium to large in size, and heavy enough to be able to throw a man to the ground by the impact of his body. Deep, powerful jaws and strong teeth are required.

155. Personnel.—*a. Function of trainer.*—Same as for trainer of the sentry dog (par. 147).

b. Function of assistant.—Same as in the case of the sentry dog (par. 147), except that the attack dog must be encouraged to bite and taught to make his bite effective. While the trainer alerts his dog to a desire to bite, it is the "fight" put up by the assistant or aggravator that determines the bite of the dog. The aggravator works under the direction of the instructor.

156. Conditions.—*a. Location.*—Attack training is initiated in the enclosed training field and completed in typical areas.

b. Time.—The attack dog must be trained to work during the day as well as at night.

c. Conditions.—Gunfire, traffic and other distractions must be used in attack dog training.

157. Equipment.—The following equipment is required:

a. Small flexible stick or other harmless weapon:

b. Chain—choke collar.

c. Six-foot leather leash.

d. Flat leather collar.

e. Stake.

f. Spring.

g. 25-foot long leash.

h. Rolled up sacking

i. Chainette or throw chain.

j. Old clothing for assistants.

k. Arm pads consisting of a padded sleeve, or sack-covered carbine boot, or any other detachable device that will protect the aggravator's arm when the dog seizes it (fig. 41).

l. Muzzles.

m. Padded undercoats for assistants.

158. Basic training required.—Same as for sentry dog training (par. 150).

159. Training procedure.—*a. General.*—The following principles and plan of attack training should be observed:

(1) *The attack dog must be taught to bite on command; he must be taught that a movement against his master constitutes a command to attack.* He must cease attack instantly on command from his master or on cessation of resistance by the aggravator. An attack dog always works with one master.

(2) *The dog must never be punished or reproved for making an attack.* Therefore it is up to the master to prevent the dog from turning on him and attacking him. He must never handle the sack or "arm" himself, nor must he expose himself to attack when the dog is unduly aroused. Similarly, the assistant who plays the part of the aggravator, or the enemy, must perfect himself in the technique of being attacked, so that he may not be injured. A green man and a green dog are liable to cause serious accidents.

FIGURE 41.—Attack training—the false arm.

(3) *Training of the attack dog passes through the following major phases:*

(*a*) The dog's natural aggressiveness is tested, aroused and controlled.

(*b*) The dog's natural tendency to attack is encouraged and his bite developed. He learns to attack on command.

(*c*) The dog is taught to transfer his attack from the sacking or other object used by the aggravator to the person of the aggravator.

(*d*) The dog is taught to cease his attack on verbal command.

(*e*) The dog learns to cease attacking when resistance ends.

(*f*) The dog learns to attack on any move which may be construed as a threat against his master, or a move which indicates an attempt to escape.

b. Specific.—The prospective attack dog first receives the training outlined for sentry work (par. 151). He is then trained as follows:

(1) *Developing the bite.*—The trainer keeps his position so that the dog, on chain and wearing the flat leather collar, is between him and the aggravator; the dog never attacks from *behind* his trainer. The aggravator presents the sacking for the dog to seize. He always offers the sacking with a sideways motion into the dog's mouth; he must not slap the dog over the face. Simultaneously, the trainer commands his dog: GET HIM. As the dog seizes the sack, the aggravator simulates a dog's growl, pulling and shaking the sack up and down. This will cause the dog to hold tightly and to fight back. After a few shakes, while the dog is holding strong and fighting vigorously, the aggravator drops the sack and runs away. The dog WINS.

(2) *Transferring the attack from the sacking and the arm pad to the aggravator.*—Without further prompting from the master, many dogs will abandon the sacking as soon as it is released, and attempt to attack the aggravator himself. In cases where the dog concentrates on the sacking and continues shaking it, disregarding the aggravator, the aggravator should turn and stamp the ground with his foot, growl, or reach in for a flank hook, in order to transfer the dog's attack from the sacking to himself. After a few days of such training the sacking is replaced by the detachable arm pad and the same procedure followed, except that the aggravator comes closer to the dog to enable him to seize the pad between wrist and elbow.

(3) *Ceasing to attack on command* OUT *and when resistance ends.*—The importance of this command cannot be overestimated, for, unless it is obeyed, the aggravator may be badly mauled or bitten. The command OUT is never repeated. It is given once and the dog must be made to obey it. The following is the procedure for teaching "out":

(*a*) The trainer has the chainette at his belt where he can reach it easily with one hand. The dog is attached by a double leash through the chain-collar ring, making a choke collar.

(*b*) The aggravator approaches. The dog seizes the "arm" and jerks it. The aggravator, instead of loosening the sleeve and giving it to the dog, continues to jerk his "arm".

(*c*) The trainer commands: OUT. The aggravator at once ceases resistance and holds his position, motionless, although the dog may continue to jerk and tug at the "arm."

(*d*) The trainer pauses for the fraction of a minute, and then jerks downward with both hands on the dog's leash. This should cause the dog to loosen his grip. In case it does not, then the trainer holds the leash tightly on one hand and throws the chainette forcefully at the dog's ribs. The moment the dog's hold is released, the trainer pulls him away from the aggravator.

(*e*) Simultaneously with the dog's releasing his hold, the command SIT is given; the dog is praised and petted as soon as he assumes the correct sitting position at the left of his trainer. Any attempt on the part of the dog to renew the attack must be checked by the command No and sharp jerks on the leash. Sufficient repetition of this exercise will teach the dog to cease attacking when resistance ends.

(4) *Attacking on leash.*—As soon as the trainer is sure that his dog has learned to attack from the chain, he teaches him to attack on leash. The training procedure is the same, except for the following:

(*a*) When the dog drops the arm pad and directs his attention to the person of the aggravator, the trainer encourages the dog to pursue the aggravator for a short distance, keeping tight hold on the leash. The aggravator falls down; the trainer checks the dog and pats him, the aggravator remaining motionless. As the dog is led away, the aggravator remains motionless until the dog is out of view.

(*b*) The trainer now walks with the dog at heel on a route along which the aggravator is hidden. As dog and trainer approach the place of concealment, the aggravator suddenly appears, simulating a growl, and presents the "arm" for the dog to seize. Simultaneously, the trainer commands: GET HIM. As soon as the dog seizes the "arm" firmly, the aggravator discards it and retreats with the dog in pursuit. The dog is allowed to catch up to him, the aggravator falls down and the trainer checks the dog with the command OUT, if necessary. This procedure is repeated in different locations (wooded areas, buildings, etc.) until the dog learns to look for the concealed aggravator, attacks him on command, and ceases the attack at the command OUT or when resistance ends. When the trainer is certain that his dog will obey him, the false arm may be abandoned. The aggravator should wear various types of clothing and uniforms over his padded undercoat.

(5) *Attacking off leash.*—In cases where a running attack must be made by the dog, he can be slipped from his leash or the leash can be dropped (fig. 42). No special training other than that on leash is required.

(6) *Attacking without verbal command.*—The dog must now be taught that a hostile, aggressive, or threatening motion on the part of the aggressor constitutes a command to attack. This must not be

attempt until the master has his dog under perfect control and is sure that he will not only attack on verbal command, but also cease his attack on command or when resistance ceases. In the new series of exercises, the master's command GET HIM is diminished in volume and tone, but always timed so as to coincide with a sudden move on the part of the aggravator. Eventually, it will be found that the command can be eliminated. The dog must be made to understand that he is not to attack unless the aggravator makes a sudden move, at which time he may attack without command.

(7) *Attacking as required under gunfire.*—Attack under gunfire is taught when the dog is well trained and the trainer is sure the dog will obey all commands or signals to attack and cease attacking. Shots are exchanged between trainer and aggravator, the first shot being fired by the trainer. Precaution must be taken not to fire close to the dog's head, since powder burns may result.

c. Supplementary.—In addition to general attack training, dogs may be taught special attack work. Particularly important is training for the following purposes:

(1) *Guarding a prisoner* (par. 18*d*).—On the command DOWN, the dog assumes this position behind a motionless "prisoner." He must be far enough away from the prisoner so that he cannot be injured by a sudden kick. The trainer then commands: WATCH HIM, which puts the dog on the alert, and walks away to a hiding place where he can observe the dog's actions. If the dog does not attack the "prisoner" when the latter attempts to turn around or run away, the trainer immediately comes out of his hiding place and commands: GET HIM. This exercise is repeated until the dog apprehends the "prisoner" on any such move. The exercise should always conclude with the trainer returning to the scene, taking charge and praising the dog. In final exercises, varying numbers of "prisoners" may be employed.

(2) *Transporting a prisoner.*—The dog is taught that a prisoner is not to be attacked when walking slowly, accompanied by the dog's master. In this exercise, the dog is made to heel on leash at the left side of the trainer, while a "prisoner" precedes both slightly to the left of the dog. The "prisoner" suddenly turns in a threatening manner, or starts to run away. The trainer releases the leash or unsnaps the dog, who, if well trained, will attack. At first, the command GET HIM may be necessary. The procedure is repeated until the dog attacks when required and refrains from attacking when the "prisoner" is marching in orderly manner ahead of the dog's master. The number of prisoners may be increased in subsequent exercises. At the conclusion of each, the dog is praised.

FIGURE 42.—Attacking off-leash.

SECTION V

DOGS FOR TACTICAL USE—THE SILENT SCOUT DOG

160. Use of silent scout dogs.—The silent scout dog is trained to detect and give silent warning of the presence of any foreign individual or group. He will prove especially useful in warning of ambushes or attempts at infiltration. The scout dog is worked by one man, the master, who has been especially trained in this type of work; he works on a long leash, in daylight or darkness, in any kind of weather, and in jungle or open country. He is aware of and gives silent warning of a foreign presence long before such presence can be detected by a human. The distance at which warning is given depends upon a number of factors: ability of the master to read his dog; wind direction and velocity; volume or concentration of human scent; humidity; density or openness of country; and amount or volume of noise or other confusing factors in the neighborhood of activity. The scout dog may be employed in various ways:

a. Reconnaissance patrols.—The patrol leader indicates the mission, the general direction of advance, and any special instructions to the master of the scout dog; he then directs the patrol to move out. The scout dog and master precede the patrol at a distance which will permit immediate communication with the patrol leader. At night this distance would probably be about an arm's length; in daylight the distance would be greater, but within easy visual signaling distance. The scout dog and his master proceed, keeping generally in the assigned direction, utilizing cover, and moving so as best to take advantage of wind and other conditions favoring the dog's power of scenting. The patrol leader may from time to time change the direction of the patrol's advance; in general, however, it is advisable to allow the scout dog and his master to move at will on the front assigned. Upon the scout dog's warning of a hostile presence, the master immediately indicates by signal "Enemy in sight," whereupon the patrol leader at once causes his patrol to take cover or halt. The patrol leader proceeds quietly, utilizing cover, to the scout dog and his master. He may order that the enemy be definitely located, in case the dog's master cannot furnish this information immediately. The scout dog and his master then proceed to work out the alert, locating the position of the enemy and informing the patrol leader. The patrol leader then takes disposition to avoid discovery, or retires, as the situation demands. He may proceed in a new direction, or circle, preceded as before by the scout dog and his master.

b. Combat patrol.—The procedure is the same as above, except that after the patrol leader has been informed of the definite location of the enemy, the scout dog and his master immediately retire to a rear or flank position so as not to interfere with any aggressive dispositions undertaken by the patrol. Upon removal of opposition, the patrol may continue its mission, preceded as before by the scout dog and his master.

c. Outposts.—In an outpost position, the main value of the scout dog is to give timely warning of approach to or attempts at infiltration through the outpost line of observation. The scout dog and his master are placed a short distance from the security group to which they are attached. This distance should be within easy visual signal in daylight, and even closer at night. A simple means of communication between the dog's master and the security group commander at night is a cord or string, which is jerked to alert the outpost. Upon being alerted, the outpost commander proceeds immediately to receive information concerning the direction and distance of the enemy. The scout dog and his master may then be directed to withdraw to a previously designated location so as not to interfere with the dispositions decided upon by the security group commander. After the situation has been cleared up, the scout dog and master are again posted as previously.

d. Leading wave preceding ground scouts.—Scout dogs together with their masters may be used to advantage to precede ground scouts in an attack. The procedure is, in the main, as defined for use with a combat patrol. However, in many instances where a dog has alerted, it will be feasible for the scout dog and his master not to withdraw, but to take cover immediately, so as least to interfere with any ensuing fire. They may continue to precede the attack when enemy dispositions are obscure or their location cannot be definitely determined. Several scout dogs working in advance on a front are able to permit attack units to advance with reasonable confidence that they will not be ambushed or unexpectedly put under fire.

e. Static security groups; combat groups; isolated positions; guarding forward dumps.—The use of the scout dog on all of the above missions is substantially as outlined for outpost duty. In all cases, local commanders should realize that the dog's master must be relied upon to advise on the use of his dog on an assigned mission. Familiarity with the powers and limitations of scout dogs come only through experience.

161. Selection of dogs for training in scout work.—*a.* To be suitable for this specialized training, dogs should possess the following traits (see pars. 14 to 18):

(1) *Intelligence.*—High.
(2) *Willingness.*—High.
(3) *Aggressiveness.*—High.
(4) *Sensitivity.*—Medium.
(5) *Energy.*—High.

b. The scout dog is basically a sharp, hardy animal of medium size. He must have highly developed scenting powers, acute hearing, and ability to detect motion. He must not be noisy or overly excitable. It is not necessary for him to possess great speed.

162. Personnel.—*a. Function of trainer.*—In addition to the functions set forth for the sentry dog trainer (par. 147), the scout dog trainer must learn to read and understand his dog so that he can interpret his every signal with regard to enemy approach or presence.

b. Function of the assistant.—The assistant is known as the aggravator or the enemy. His functions are the same as those of the assistant in sentry dog training (par. 147).

163. Conditions.—*a. Location.*—Training starts in the regular training area. It is inadvisable, at first, for the "enemy" to conceal himself behind trees, rocks, bushes, or any other screen. If the dog, on his initial scouting assignment, finds the "enemy" concealed behind some natural cover, he will learn to associate such spots with his discovery of the enemy, and will consequently rely on his inferior eyesight to discover him. He will not use his nose and ears, which are the senses he must learn to depend upon. Training locations must be changed daily, so that the dog does not learn to associate the "enemy" with a given area.

b. Hours.—Although a scout dog is generally used at night, he must be trained during the daytime. The instructor decides when a dog and his trainer have completed training and are ready for exercise at night. Training is not done at night because the dog's behavior cannot be observed closely enough.

c. Wind.—Although a breeze is desirable at all times during training, it is not absolutely essential.

164. Equipment.—The following equipment is required:

a. Small flexible stick or other harmless weapon.
b. Chain-choke collar.
c. Six-foot leather leash.

d. Flat leather collar (special for scout training).

e. Stake.

f. Spring.

g. Harness (special for scout training).

h. 25-foot long leash.

i. Rolled-up sacking.

j. Chainette, or throw chain.

165. Basic training required.—Same as for sentry dog training (par. 150).

166. Training procedure (figs. 43 to 46).—*a*. Training starts and proceeds through the steps outlined for sentry dog training (par. 151),

FIGURE 43.—Silent scout dogs winds an enemy. The first indication of suspicion.

FIGURE 44.—Silent scout dog indicates the enemy's hiding place by pointing.

FIGURE 45.—Silent scout dog takes his trainer to the enemy's place of concealment. (Note raised hackles on dog's neck and back.)

except that *the dog is reproved if he attempts to bark as the enemy appears with his sacking.* The trainer must use every means to quiet the dog. If the dog is taught to understand the correction "No" in basic training, it is usually not difficult to quiet him thereby. The dog's mouth may be kept shut gently with the hand. As soon as the dog is quiet or is content to growl very softly, he should be patted and praised.

b. The dog and his trainer next proceed into an open field, preferably one where high grass affords cover for the "enemy." Once in

FIGURE 46.—Silent scout dog routs the enemy.

the field the dog is put on the 25-foot leash and equipped with the special collar *which is used only when the dog is working, and removed at the conclusion of the exercise* (fig 47). The "enemy" should be concealed upwind from the dog so that his scent is driven directly towards the dog's nose. The trainer commands: WATCH HIM. The dog will then attempt to locate the "enemy" by scent, ear, or sight. It is important that for this initial exercise the "enemy," although well concealed from vision, should be within easy scenting distance and directly upwind from the dog.

c. The dog is encouraged to lift his nose and sample the wind well above ground level. When he detects a scent other than his trainer's, he is likely to react in one of the following ways:

(1) General tenseness of the whole body.

(2) Hackles raised.

(3) Ears pricked.

(4) Other forms of alertness easily recognizable by a keen observer, such as keenness to investigate, slight inclination to whimper or growl, tail active or distinctly rigid.

FIGURE 47.—The silent scout dog. (Trainer removes chain choke collar and adjusts special flat leather collar prior to initiating training.)

d. As soon as he gives such evidence that he has picked the scent, he is praised. Praise at this point must be given in a soft, almost whispering voice; it must never be insistent enough to divert the dog from his work. (Some dogs when first going into the field will put their noses to the ground and attempt to pick up a ground scent. This tendency must be immediately discouraged by the trainer who should raise his foot under the dog's chin and utter the reproof "No, no" in a stern voice.)

e. When the dog has winded the "enemy" and approximated the direction of his hiding place, the "enemy" appears, waves the sacking at the dog, and runs away. The trainer rewards the dog and the exercise is considered concluded.

f. During repetitions of the above procedure, the distances between the dog and the concealed "enemy" are lengthened progressively. The trainer must give his dog every possible opportunity to locate the "enemy." He stops at intervals to take advantage of every breeze, *quartering the field if necessary to catch the scent* (fig. 48). He pays

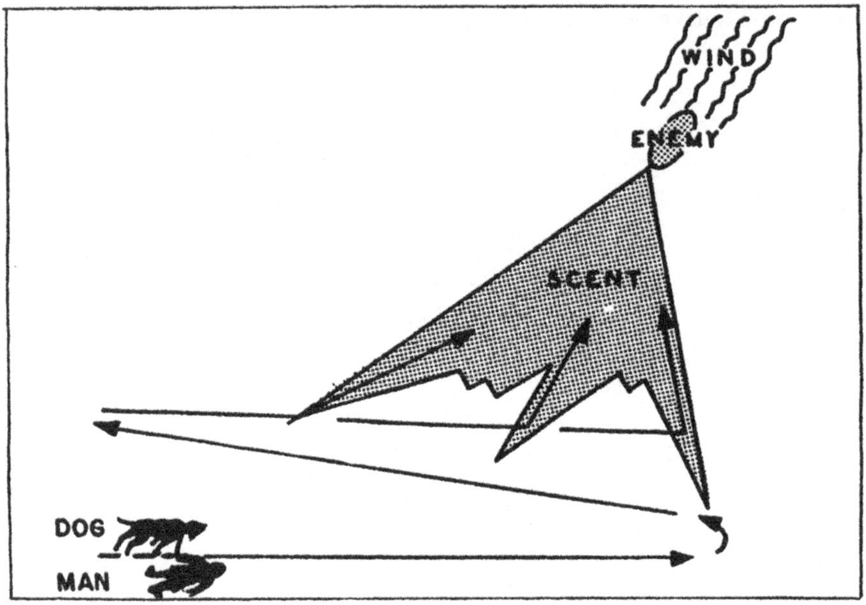

FIGURE 48.—Route of the master and silent scout dog in quartering a given area. Final arrows indicate points at which the dog picks up the "enemy's" scent.

strict attention to the dog's demeanor and encourages the dog as soon as he shows signs of suspicion.

g. This procedure is repeated daily with the conditions varied. The "enemy" is concealed in underbrush, in ditches, in the branches of a tree, behind rocks or in any natural or artificial hiding place.

h. The role of the "enemy" is played by a different man every day so that the dog will learn to pick up any human scent other than his master's. More than one "enemy" may be used simultaneously for a given exercise. Occasionally the dog is allowed to rout and pursue the "enemy" or "enemies" in order to maintain his alertness.

i. After the dog and his master have established mutual understanding, it will not be difficult to locate the "enemy" at night; most dogs are much keener at night.

167. Training procedure for stationary work.—The training procedure for use of a dog on an outpost as a precaution against enemy infiltration differs from the foregoing in that the dog remains motionless and the concealed "enemy" advances. The dog is trained to alert while stationary. His master judges the "enemy's" approach in the light of the dog's demeanor.

SECTION VI

DOGS FOR TACTICAL USE—THE MESSENGER DOG

168. Use of messenger dog.—*a.* Wherever the use of a soldier runner is indicated, a messenger dog should be used instead. He is surer and faster; he can find his way in daylight or darkness, in any kind of weather, over rough or smooth terrain, open or jungle country, at high altitude, and in snow and cold. He can carry a message up to 1 mile at great speed. He is a difficult target, due to size, speed, and natural ability to take advantage of cover (fig. 49). The use of a messenger dog in place of a runner will not only insure a more rapid and reliable means of communication, but may also save human life and limb.

b. Messenger dogs may be used in connection with scout dogs on reconnaissance patrols, combat patrols, outpost, static security activities, or lines in observation. In addition, they may be used to establish communication between two fixed centers; between a fixed and moving center; or between two moving centers; for packing limited emergency supplies (in an *emergency* messenger dogs may be relied upon to carry not in excess of 25 pounds); and as a quick means of laying wires over short distances. Carrier pigeons, inclosed in special carriers, may be transported by messenger dog.

169. Selection of dogs for training in messenger work.—*a.* To be suitable for this specialized training, a dog should possess the following traits (see pars. 14 to 18):

(1) *Intelligence.*—High.

(2) *Willingness.*—High.

(3) *Energy.*—High.

(4) *Aggressiveness.*—Moderate.

(5) *Sensitivity.*—Moderate.

b. The messenger dog accomplishes his mission *motivated solely by* his eagerness to please his two masters. Consequently, a potential messenger dog must show marked affection and loyalty for man. He must have great courage, for on many occasions he must carry his message through gun and shell fire. He must have stamina, strength, and endurance to carry him through difficult terrain. He must possess superior powers of scent and hearing, great speed, and the ability to swim. He should be of medium size, large enough to negotiate difficult terrain. The messenger dog, unlike the sentry,

FIGURE 49.—The messenger dog is a difficult target.

scout, or attack dog, must not look for trouble. He must try to evade capture, but if he is cornered, he must be able to defend himself. He must be of a suspicious, rather than an aggressive nature, trusting and obeying only his two masters. He must never respond to advances made by any other individual.

170. Personnel.—Two trainers are assigned to each potential messenger dog. They work to divide the dog's affection and interest equally between them. They take equal turns feeding and exercising the dog. They share equally in training exercises. They take turns in allowing the dog to perform exercises which he likes and does well.

The instructor makes it his special duty to insure that the two men acting as masters of a dog are so matched that they can work with each other and achieve equal places in the dog's affections. Assistants may be used to watch and correct the dog when he is on a mission and out of sight of the two masters.

171. Conditions.—*a. Location.*—Initial training takes place in the enclosed training area. It is completed over unfamiliar terrain presenting varieties of obstacles.

b. Hours.—Training must be completed during daylight but final tests must be made in darkness. (See par 163*b*.)

c. Circumstances.—Every known distraction must be introduced during training, such as bomb detonations, shell and gunfire, troop and motor traffic, and the presence of moving or stationary troops, civilians, and animals.

172. Equipment.—The following equipment is required:

a. Chain choke collar.

b. 6-foot leather leash.

c. 25-foot leash.

d. Special message collar with pouch attached.

e. Chainette.

173. Basic training required.—See paragraph 150.

174. Training procedure.—*a. General.*—Training of the messenger dog is based on the affection of the dog for his two trainers. Both trainers and the dog should have ample time to become acquainted with each other. The trainers must play with the dog, feed and pet him, and otherwise win his affection. They must learn and employ a uniform pronunciation for commands, the same gestures, and the same methods of praising and rewarding. The two trainers take turns working the dog in basic training for at least 30 minutes every day; they take the dog for exercise and for walks together. Each dog's affection may be won by different methods, but it will not take the two trainers long to find out what particular form of reward appeals most to the particular dog in their charge. It usually takes about 2 weeks for the dog to recognize his masters, to realize they are his friends, and to disregard every other individual. When this is established, specific training may proceed.

b. Specific.—(1) The two masters, henceforth designated as *X* and *Y*, take the dog to an inclosed training area and stand about 20 feet apart, facing each other. *X* holds the dog's leash, plays with him and pets him. He then commands the dog: SIT, drops to one knee, and attaches the special message-carrying collar around his neck. *X* commands: REPORT, gesturing outward with his arm and dropping the leash. Simultaneously, *Y* drops to one knee, calls the dog, and when he comes to him, commands: SIT while he removes the message-carrying collar. When the collar is removed, the trainer rises and praises the dog.

(2) As soon as the dog runs unhesitatingly from *X* to *Y* and sits in front of *Y* when he reaches him, the same procedure is repeated, with *X* calling the dog from *Y* and removing the collar when the dog reaches him. It must be borne in mind that the special collar is put on the dog just before he runs from one master to the other, and removed as soon as he completes the exercise; thus the dog learns to associate the collar with the need to go from one master to the other. Further-

FIGURE 50.—Dispatching the messenger dog.

more, the master is always in a kneeling position when he dispatches or receives the dog (figs. 50 and 51). This helps the dog to learn to ignore people walking or standing about.

(3) The distance between the two masters is extended until a minimum of 100 yards is reached, both masters remaining in full sight of the dog, calling and encouraging him alternately if he stops en route.

(4) When the dog runs at good speed between his two masters, concealment is introduced. *X* and *Y* walk a short distance together, *X* holding the dog's leash. *Y* walks about 50 yards ahead and then seeks a hiding place (underbrush, trees, a house, a trench, or ditch, tall grass, etc.). *X* kneels, puts the message collar on the dog, and comands: REPORT! with outflung arm. If the dog runs at once to find *Y*, it will not be necessary for *Y* to call the dog. However, in case the dog hesitates before proceeding, or, having started, is

unsuccessful in finding *Y*, *Y* must reveal his presence and, if necessary, call the dog. The distance between the two masters at first is limited so that it may be relatively easy for the dog to find *Y*. The dog must always *be successful in his search; he must never return without having found the hidden master.* Therefore, if he has difficulty finding him the first time, the next exercise should be made easier. Thereafter, difficulty may be progressively increased. *The same hiding place must never be used twice. The dog must search or learn to use his nose in finding his master.* As the dog becomes more certain in his mission, the distance between the two masters is increased until it extends to about 1 mile.

FIGURE 51.—Receiving the messenger dog.

(5) The two masters now alternate sending and receiving the dog, with the training area changed as often as possible.

(6) The dog now learns to return to his dispatching master though the latter has moved away from his original position. *X* sends the dog with a message and then takes up a new position near to his original one. *Y* sends the dog back in the now usual manner. The dog should experience no difficulty in finding *X*. The distance from the original point may then be progressively increased and places of concealment introduced and rendered progressively difficult to discover. Finally, the number of round trips (carrying message and returning with reply) may be varied, and one-way trips alternated.

175. Additional training notes.—*a. Swimming.*—Messenger dogs must be taught to enter water and swim, if such will shorten the line of communication. The first lessons should be in water (a slow stream, if possible) which the master can wade but which the dog must swim; it should be wide enough to keep the dog from jumping across. If the dog hesitates to enter the water, he must not be thrown in; the master enters the water and then with a slight pull on the leash calls the dog.

b. Distractions.—When the dog is working well between his two masters, distractions must be introduced gradually so that the dog learns to run between his two masters, regardless of gunfire, traffic, or any other distraction that may be prevalent.

c. Use of artificial scent.—In certain situations the use of artificial scent may be employed to advantage. This is accomplished as follows: a small piece of cloth is impregnated with aniseed or any other holding scent. The cloth is weighted and dragged at the end of a string by master *X* as he leads his dog. The dog is thereby aided in finding his way back to the point of departure where he may be relieved of the message he is carrying. On his return journey to *X*, the dog can pick up the scent again and readily find *X*, who has continued to drag the scent while advancing to a new position. The chief disadvantages of artificial scent are the possibility of confusion where different scent trails have crossed, and the likelihood that dogs following artificial scent may come to rely upon it exclusively.

d. Extension of training.—Dogs assigned to using units must receive continual training practice in order to be kept up to standard. This practice should be daily if the military situation permits. Daily training should never include more than two round trips each morning and each afternoon. Considerable time must be allowed between each trip for the dog to rest. When night work is undertaken, one training period must be omitted during the day. Training should also include exercises in carrying messages after long marches.

e. Changing masters.—When conditions render it necessary to change masters, they must never be replaced simultaneously. One master must remain to train the replacement. In general, masters should be replaced *only when necessary;* it is desirable to have the same two masters continue with the dog's feeding, grooming, exercise, and daily training. *Under no circumstances should anyone other than the two masters be allowed to handle or play with the dog.*

SECTION VII

DOGS FOR TACTICAL USE—THE CASUALTY DOG

176. Use of casualty dogs.—The casualty dog is trained and used to aid the medical corps in locating wounded on battlefields and in other areas. Before losing consciousness, injured soldiers may crawl for safety to hiding places that may easily be overlooked by patrolling medical units; they may be buried in debris caused by bombings. It is obvious that many lives can be saved by prompt treatment, especially in cases of shock or hemorrhage, and it is the function of the casualty dog to discover such injured, to report his discovery to his master, and to lead help to the casualty. The dog is trained to search a given area and, upon discovering a casualty, to return to his master and report his find. The master then fastens a leash to the dog's collar, and is led by him to the casualty.

177. Selection of dogs for training in casualty work.—*a.* To be suitable for this specialized training, dogs should possess the following traits (see pars. 14 to 18):

(1) *Intelligence.*—High.
(2) *Willingness.*—High.
(3) *Energy.*—High.
(4) *Aggressiveness.*—Low.
(5) *Sensitivity.*—Moderate.

b. The casualty dog must also possess superior power of scent, endurance, good sense of direction, and keen hearing. He must not be noisy, excitable, or inclined to fight with other dogs. He is medium in size. Underaggressiveness is a particularly important trait. Delirious wounded may strike at an approaching dog; the dog must be relied upon not to bite and inflict additional injury upon a casualty, even in self-defense.

178. Personnel.—*a. Function of trainer.*—The function of the trainer is to teach the dog to find, report, and lead to a casualty.

b. Function of the assistant.—The primary function of the assistant is to act as a casualty for the dog to discover. In performing this function, the assistant must observe certain precautions that are necessary to insure effective training.

179. Conditions for training.—*a. Location.*—Training starts in the regular training area. As training progresses, the location is varied to include conditions most likely to be encountered in a theater of operations.

b. Hours.—Initial training is undertaken during the daytime. When proficiency warrants, night exercises are carried out.

180. Equipment.—The equipment necessary for training in casualty work, in addition to that used for basic exercises, is as follows:

a. A 25-foot training leather, rope or tape leash.

b. Special training harness.

c. Collar bearing Red Cross insignia (figure 52).

FIGURE 52.—Special collar for casualty dog.

181. Basic training required.—*a.* Specialized training in casualty work is based on training in the following exercises, which are described in detail in chapter 3:

(1) Heel (close, loose).
(2) Standing halt.
(3) Stay.
(4) Come.
(5) Cover.
(6) Jump.
(7) Crawl.

b. Dogs must be able to perform the above exercises undisturbed by gunfire or other distractions.

182. Training procedure.—*a. General.*—(1) The training of the casualty dog embraces three major functions:

(*a*) *Searching.*—The dog ranges a given area, searching and finding a casualty (fig. 53).

(*b*) *Reporting.*—The dog returns to his master, informing him by his manner that he has discovered a casualty (fig. 54).

FIGURE 53.—Searching. Discovery of the casualty.

FIGURE 54.—Reporting. Reports the discovery of a casualty.

(*c*) *Leading.*—The dog, on leash, leads his master to his discovery (fig. 55).

(2) The dog is taught these functions in the order indicated. However, as training progresses, the dog is made to learn that they constitute a single mission. The trainer never permits one of the functions to take precedence over either of the other two in working the dog. Thus, if the dog acquires greater proficiency in searching, he must be given additional time and attention in reporting and leading,

FIGURE 55.—Leading. The dog leads his master to the casualty. (Note eagerness and drive in dog's attitude.)

so that all three functions are equalized. The dog's interest must be kept at the same level for all three functions throughout training.

(3) The harness is always adjusted just before the dog is given the order to search, and removed just after he has led his master to the casualty.

(4) The trainer always encourages the dog to lead as vigorously as possible and refrains from holding back the dog.

(5) The dog is always given a rest period after he has accomplished his mission.

b. *Specific.*—(1) Before training for the specific functions of searching, reporting, and leading is initiated, the dog is taught to heel without verbal command; he is also taught to use his native ingenuity in surmounting obstacles.

(2) The trainer adjusts the harness on the dog. The assistant or casualty lies down at a distance of 5 or 6 feet from the trainer, encouraging the dog to come to him by gestures. When the dog obeys, he praises and pets him. The trainer joins the assistant and adds his praise, simultaneously grasping the dog's harness (fig. 56).

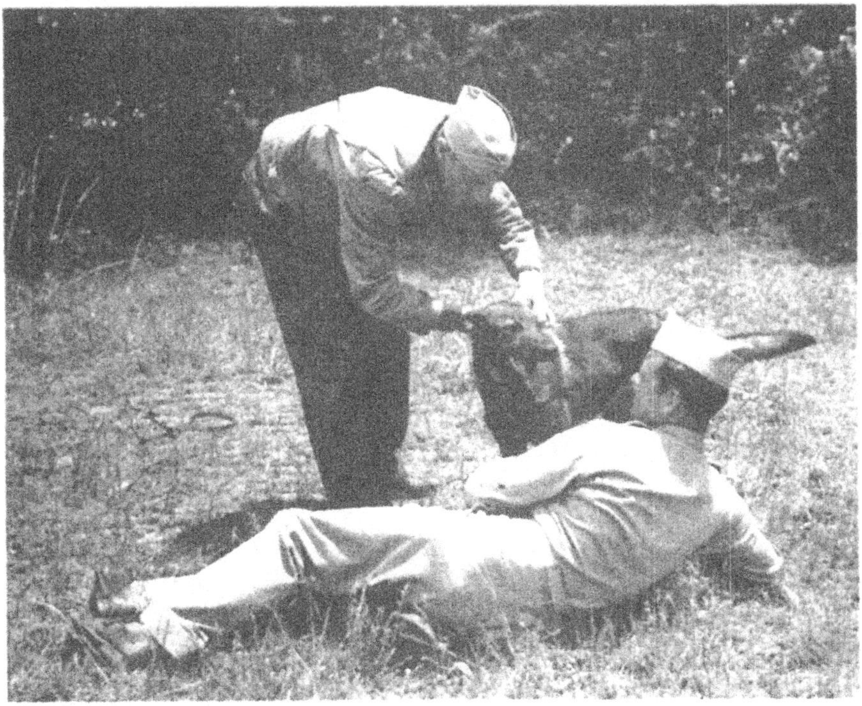

FIGURE 56.—The casualty dog. Trainer and casualty simultaneously praise the dog.

(3) The casualty rises quickly and proceeds for about fifty yards at a stooping run (fig. 57). He drops to the ground again. The trainer, holding the harness tightly, encourages the dog to follow him. He gives the command: SEARCH with the "S" accented so that the word is almost hissed (fig. 58). The casualty falls, the trainer sends the dog after him, releasing the harness with an outward gesture, repeating the command SEARCH and again dwelling on the "S" sound. He follows the dog as soon as he reaches the casualty. The casualty and the trainer jointly praises and reward the dog.

(4) The trainer backs away from the casualty; when he has moved

2 or 3 yards away, he stops and calls the dog, praising and petting him when he comes (fig. 59). The casualty immediately rolls a few yards away. The trainer holds the harness until the dog's attention is attracted by the casualty. He then releases the dog without command, this time following the dog at once. When the dog reaches the casualty, both men praise and caress the dog simultaneously. The trainer then removes the dog's harness and returns to the starting point with the dog off leash and under no restraint (fig. 60). The

FIGURE 57.—The casualty proceeds at a stooping run to attract dog's attention.

casualty rises, making sure that the dog is not observing him, and proceeds alone by another route to the starting point. This procedure is repeated, not more than four times daily, allowing the dog to rest between exercises, over a period covering 1 to 7 days. The length of this period is determined by the dog's progress. It is advisable to kennel or chain the dog during rest periods, as this makes him more eager to work.

(5) The casualty dashes off, disappearing from the dog's sight and continues for some distance beyond that point before lying down. The procedure described in (4) above is then followed.

(6) The casualty leaves the starting point, unobserved by the dog, and hides.

(7) The trainer adjusts the harness on the dog, encouraging him to search. At this stage of training the dog will probably be eager to perform his task. If he does not evidence this eagerness, he is further urged by the hissed SEARCH command, and the procedures in (3) and (6) above are repeated until the dog shows spirit and eagerness to search. The command is eventually eliminated.

(8) As soon as the trainer is satisfied with the dog's eagerness to be off and his speed and vigor, when released, the next stage of his

FIGURE 58.—The casualty dog. The command search.

training is undertaken. The casualty is placed 20 to 30 yards from the trainer and the dog. He remains motionless when the dog discovers him. The dog behaves, according to his temperament, in one of the following ways, and training is continued accordingly:

(*a*) He displays great interest in the casualty, sniffing at him and nudging at him with his nose. In this case the casualty taps the dog lightly on the muzzle, using as little force as possible and limiting his motions. The trainer attracts the dog's attention and when the dog looks to him for direction, the trainer backs up rapidly.

(*b*) The dog, after sniffing briefly at the casualty, raises his head and looks at his trainer for direction. The trainer runs backward, calling and encouraging the dog; when he comes, the leash is attached to the harness. The casualty moves away on all fours, the trainer urging his dog to lead him to the casualty. When the dog has done this, the trainer and casualty praise him simultaneously. The trainer and casualty will use their judgment in furnishing incentive for the dog to work between them. Encouragement should vary in ratio to the dog's understanding of his mission, to the eagerness and vigor he displays in searching, reporting, and leading.

FIGURE 59.—The casualty dog. The trainer backs up and encourages the dog to him.

(9) As the dog progresses in this training, the casualty is found in increasingly difficult locations. A second casualty may be introduced.

(10) Other variations are introduced, where practicable:

(*a*) Clothing, sacking, etc. are placed within the training area.

(*b*) A second assistant stands or sits within the area, close to the starting point.

(11) *Final training.*—(*a*) It is essential that battle conditions be simulated, since these are the conditions under which the dog will have to work eventually. Gunfire is introduced. Casualties are covered with coats or bandages; if practical, their clothing should appear

to be blood-stained. Trees, shell-holes, trenches, and fox holes are utilized for casualty locations. Casualty dogs are taught to disregard scattered humans who are standing, sitting, or walking. Finally, night work is introduced and continued until proficiency is attained under difficult conditions in darkness.

FIGURE 60.—The casualty dog. After completing the exercise, the dog returns to the starting point, off-leash.

(*b*) As training progresses, the training harness is discarded and the longer leash is looped to enable the dog to lead his trainer to the casualty. The harness is not used in actual service because of the possibility of its catching on branches, wire fences, and similar objects.

183. Using principles.—The competent master never allows his dog's interest to wane. Constant fruitless searches will cause the dog to lose his eagerness; yet in actual service he will often operate in areas where there are no casualties. After each negative report, the master will give his dog the customary rest period and then arrange a search area with simulated "casualties." The dog, upon being sent out this time, will inevitably find, report, and lead to the simulated casualty. The satisfaction the dog derives from this success will obliterate his prior disappointment, thus keeping his interest aroused.

Section VIII

USING PRINCIPLES

184. General.—Throughout chapters 3 and 4, reference has been made to certain principles upon which will depend the success of the training program. Certain considerations are no less important in their application to the actual use of war dogs. These include—

a. Dogs assigned to using units must receive continual practice or refresher training to insure a high level of performance.

b. Masters of war dogs should be replaced only when necessary for strategic reasons.

c. The effectiveness of war dogs depends to a significant extent on the regard shown for their well-being: On proper grooming, feeding, and kenneling, within the limits determined by the tactical situation.

d. It must be understood by troops coming in contact with war dogs that they are not to be considered as pets or mascots.

e. War dogs are not to be handled by anyone except their masters.

f. Whenever possible, masters of war dogs should be notified several hours in advance of contemplated activity, so as to insure proper preparations.

g. War dogs must never be fed immediately before a scheduled mission.

h. Masters should be consulted beforehand to determine whether or not prevailing conditions are suitable for employment of war dogs to accomplish desired objectives.

APPENDIX I

TRAINING PROGRAMS

This manual has described methods to be followed in training war dogs and their masters. Training should be standardized as far as possible along the lines indicated. Because conditions vary considerably, it is not feasible to provide in this manual a standard detailed schedule nor to specify the number of hours to be devoted to each phase of instruction. Adjustments have to be made in accordance with the number of men and dogs to be trained, their quality, the number and qualifications of the instructional staff, and facilities and time available. Under favorable circumstances it is expected that training of war dogs and, simultaneously, of their masters can be accomplished within the time periods indicated below:

Class	Weeks
Sentry	8
Attack	13
Scout	13
Messenger	13
Casualty	13

It is to be borne in mind that in addition to practical work with their dogs, trainers must be given lectures and demonstrations based on the content of all parts of this manual. Demonstrations pertinent to a given practical exercise must in all cases precede the actual exercise. Lectures, however, may be given when convenient; for this purpose advantage may be taken of inclement weather. In general, instructions in grooming, feeding, and kenneling of dogs must come early in the program since the success with which men train dogs will depend to a definite extent on the care they give their dogs.

The chart below illustrates an effective plan for training scout dogs and masters under favorable conditions. This chart may be used as a guide in setting up plans for any branch of training, with due adjustments for prevailing conditions (supply of and demand for men and dogs; instructional staff; facilities; time).

Each class completes its training within 13 weeks. During the first 10 weeks the class is in the hands of the instructor and of five assistant instructors, each for a 2-week period. The first assistant provides 2 weeks of basic training; the second, 2 weeks of further basic and specialized training; the third, fourth, and fifth, 2 weeks each

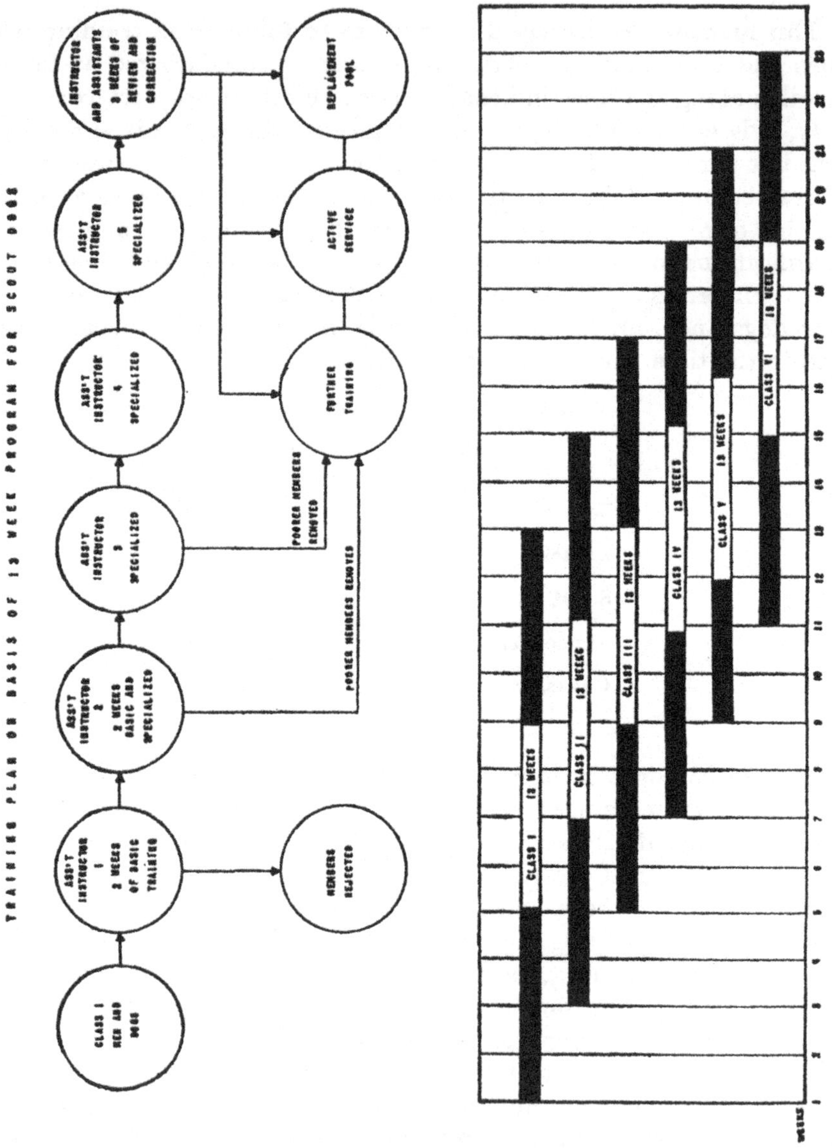

of specialized training. During the first 2 weeks of training some members of the class may be found entirely unsuitable for the work and are eliminated. As training progresses, slow members of the class may be removed for individualized training or assignment to a subsequent class. In the final 3 weeks, the class undergoes reviews and remedial instruction under the direction of the chief instructor with the assistance of staff members. This procedure provides an excellent means of determining readiness for service. Some members of the class may be immediately assigned to using agencies, some to a replacement pool; others may be deemed to be in need of additional training and are assigned accordingly.

The plan illustrated is based on the assumption that a new class of men and dogs will enter the training center every 2 weeks. Thus, when assistant instructor No. 1 has completed his work with class No. 1, he is ready to take charge of class No. 2, entering 2 weeks after class No. 1. The plan is one of rotation and provides a means whereby every class receives the benefit of the experience and ability of every instructor attached to that particular branch of war dog training.

Appendix II

BIBLIOGRAPHY

The following sources have been consulted in the preparation of text material for this manual:

The Complete Dog Book (1941). Published by the American Kennel Club, New York, N. Y.

War Dogs of Other Days by M. V. Morden (American Kennel Gazette, October 1942).

War, Police, and Watch Dogs by E. H. Richardson (1910). Published by William Blackwood & Sons, London.

Working Dogs by Elliott Humphrey and Lucien Warener (1934). Published by the John Hopkins Press, Baltimore, Md.

Manuscript materials in the Office of The Quartermaster General (Remount Branch), War Department, Washington, D. C.

Companion Dog Training by Hans Tossuti (1942). Published by Orange Judd Publishing Company, New York, N. Y.

INDEX

[A. G. 300.7 (9 Jun 43).]

BY ORDER OF THE SECRETARY OF WAR:

G. C. MARSHALL,
Chief of Staff.

OFFICIAL:

J. A. ULIO,
Major General,
The Adjutant General.

DISTRIBUTION:

X.

(For explanation of symbols see FM 21–6.)

Made in the USA
Monee, IL
05 November 2026

67949850R00090